CONTENTS

Easy

GROCERY BAG
page 6

BOTTLE BAG
page 7

GIFT BAGS TO KNIT AND CROCHET
page 8

TEA TIME PLACEMAT
page 10

KITCHEN RUG
page 11

KITCHEN BASKETS
page 12

KITCHEN ORGANIZER
page 14

SPRINGTIME TRIANGLE
page 17

PRETTY PASTELS
page 18

PRIMARY COLORS DISHCLOTH
page 19

DISHCLOTH AND SCRUBBY
page 20

SPRING DREAMS DISHCLOTH
page 21

WITH A TWIST
page 22

LACY DISHCLOTHS
page 24

SPA SET
page 28

SPA BAG
page 29

DRAWER ORGANIZER
page 30

BATH BOXES
page 31

SPA CONTAINERS
page 33

BEACH BAG
page 35

CHECKERED PICNIC BLANKET
page 37

TRICK OR TREAT
page 40

AUTUMN HARVEST
page 42

SEASONS GREETINGS
page 43

MERRY CHRISTMAS
page 44

page 8

Beginner

page 26

SPA FACE CLOTH
page 26

PASTEL WASHCLOTHS
page 27

TISSUE BOX COVER
page 32

4TH OF JULY
page 34

Intermediate

MOM AND ME
page 2

SUSHI SET
page 4

LOOPY DISHCLOTH
page 16

STARS AND STRIPES CUSHION
page 36

OVEN MITTS
page 38

FLAG APRON
page 39

pages 38-39

MOM and me

YOU'LL NEED

YARN (4)
Lily® Sugar'n Cream®
Solids 2.5oz [70.9g], 120yds [110m]

Main Color (MC)
00026 (Light Blue)
Child: 2 balls
Adult: 5 balls

Contrast (A)
00007 (Ivory)
Child: 2 balls
Adult: 4 balls

HOOK
Size H (5.0mm) crochet hook *or size to obtain gauge*

MEASUREMENTS
Child size Approx 17 x 20" [43 x 51cm].
Adult size Approx 22½ x 23" [57 x 58.5cm].

GAUGE
13 sc and 14 rows = 4" [10cm]. *Remember to check your gauge!*

NOTES
1 The instructions are written for smaller size. If changes are necessary for larger size the instructions will be written thus ( ). When only one number is given, it applies to both sizes.
2 To change color, work to last 2 loops on hook at end of row. With new color, draw through 2 loops on hook, then proceed in new color.

APRON
With MC and larger hook, ch 56 (74).
1st row (RS) 1 hdc in 3rd ch from hook (counts as 1 hdc). 1 hdc in each ch to end of ch. Turn—54 (72) hdc.
2nd row Ch 2 (does not count as st). 1 hdc in each hdc to end of row. Join A. Turn.
3rd and 4th rows With A, as 2nd row, joining MC at end of last row.
5th and 6th rows With MC, as 2nd row, joining A at end of last row.
Rep 3rd to 6th rows for Stripe Pat until work from beg measures 12" (15") [30.5 (38)cm], ending with a WS row.
Next row With appropriate color, ch 2. (Yoh and draw up a loop in next st) twice. Yoh and draw through all loops on hook - Hdc2tog made. 1 hdc in each st to last 3 sts. Hdc2tog over next 2 sts. 1 hdc in last st. Turn. Place markers at each end of row. Keeping cont of Stripe Pat, rep last row until 24 (28) sts rem.
Cont even in Stripe Pat until work from beg measures approx 19" (22") [48 (56)cm], ending on a 6th row of Stripe Pat. Fasten off.

Pocket
With MC, ch 56 (74).
1st row (RS) 1 hdc in 3rd ch from hook (counts as 1 hdc). 1 hdc in each ch to end of ch. Turn—54 (72) hdc.
2nd row Ch 2 (does not count as st). 1 hdc in each hdc to end of row. Turn.
Rep last row until work from beg measures 5½ (7½)" [14 (19)cm], ending with a WS row. Fasten off.

Vertical Dividers Place vertical markers along 19th (25th) hdc in from each side edge.
With RS facing, join 2 strands of A with sl st at foundation edge of marker and work a vertical chain up length of Pocket to top edge. Fasten off.
Rep along other marker.

Bottom edging
1st row With RS facing, join A with sl st in bottom corner of lower edge. Ch 1. Work 54 (72) sc evenly across. Fasten off.
2nd row With RS facing, join MC with sl st in first sc of last row. Ch 1. Work 1 sc in each sc across. Fasten off.

Top edging Work as for Bottom edging across top edge.

Pin Pocket in position to RS of Apron, having bottom edge of Pocket 2 (3)" [5 (7.5cm) above lower edge of Apron.

Apron Edging
1st rnd With RS facing, join A with sl st to bottom right corner of Apron. Ch 1. 3 sc in same sp as sl st. Work 53 (71) sc along bottom edge, 3 sc in corner, 38 (49) sc up side edge (working through both Apron and Pocket edges), 22 (32) sc up armhole edge,

3 sc in corner, 23 (27) sc across top edge, 3 sc in corner, 22 (32) sc down armhole edge, 38 (49) sc down side edge (working through both Apron and Pocket edges). Join with sl st in first sc.
2nd rnd Ch 1. 3 sc in same sp as last sl st. Work 1 sc in each sc around, having 3 sc in each corner sc. Join with sl st in first sc. Fasten off.

Neck strap
With A, ch 58 (63).
1st row 1 sc in 2nd ch from hook. 1 sc in each ch to end of ch. Turn—57 (62) sc.
2nd and 3rd rows Ch 1. 1 sc in each sc to end of row. Turn. Fasten off at end of 3rd row.

Waist Ties (make 2)
With A, ch 85 (101).
1st row 1 sc in 2nd ch from hook. 1 sc in each ch to end of ch. Fasten off.

FINISHING
Sew Pocket vertical dividers through both thicknesses to form 3 pockets. Sew Neck Strap and Waist Ties in position.

SUSHI set

YOU'LL NEED

YARN (4)
Lily® Sugar'n Cream®
Solids 2.5oz [70.9g], 120yds [110m]

Contrast A
2 balls 50300 (Plum)

Contrast B
2 balls 50610 (Gold)

Contrast C
2 balls 01742 (Hot Blue)

Note Amounts are given for 3 Placemats and 3 Napkin Rings (one of each color combination)

HOOKS
Placemats Size H (5.0mm) crochet hook *or size to obtain gauge*

Napkin Rings Size G (4.0mm) crochet hook *or size to obtain gauge*

MEASUREMENTS
Placemats Approx 13" x18" [33 x 45.5cm].
Napkin Ring Flower approx 2½" [6cm] in diameter.

GAUGES
15 sc and 16 rows = 4" [10cm] with smaller crochet hook.
13 sc and 14 rows = 4" [10cm] with larger crochet hook. *Remember to check your gauges!*

Note To join new color, work to last 2 loops on hook. Draw new color through last 2 loops then proceed in new color.

PLACEMAT
With Color 1, ch 50.

1st row (RS) 1 sc in 2nd ch from hook. *1 dc in next ch. 1 sc in next ch. Rep from * to end of ch—49 sts. Turn.
2nd row Ch 3 (counts as dc). *1 sc in next dc. 1 dc in next sc. Rep from * to end of row. Turn.
3rd row Ch 1. 1 sc in first dc. *1 dc in next sc. 1 sc in next dc. Rep from * to end of row. Turn.
Rep 2nd and 3rd rows for Pat until work from beg measures approx 10" [25.5cm], ending with a WS row. Fasten off.

Border 1st rnd With RS of work facing, join Color 2 with sl st to top right corner. Ch 1. 3 sc in corner. Work 48 sc across to next corner. 3 sc in corner. Work 31 sc down left side to next corner. 3 sc in corner. Work 48 sc across bottom to next corner. 3 sc in corner. Work 31 sc up right side to next corner. Join with sl st to first sc. Turn.
2nd rnd Ch 1. Working in back loops only, 1 sc in each sc, working 3 sc in corner sc. Join with sl st to first sc. Turn.
Rep last rnd until border from 1st rnd measures 1½" [4cm], ending with a RS rnd. Fasten off.

Make 1 Placemat having A as Color 1, C as Color 2.
Make 1 Placemat having B as Color 1, A as Color 2.
Make 1 Placemat having C as Color 1, B as Color 2.

Chopsticks Holder (make 3)
With Color 1, ch 8.
Work in pat as given for Placemat for 2" [5cm], ending with a WS row. Fasten off.

FINISHING
Center Chopsticks Holder at right side edge on inner side of Border and sew sides in position, leaving top and bottom edges of Holder free.

NAPKIN RING
With Color 1, ch 10.
1st row (RS) 1 sc in 2nd ch from hook. *1 dc in next ch. 1 sc in next ch. Rep from * to end of ch—9 sts. Turn.
2nd row Ch 3 (counts as dc). *1 sc in next dc. 1 dc in next sc. Rep from * to end of row. Turn.
3rd row Ch 1. 1 sc in first dc. *1 dc in next sc. 1 sc in next dc. Rep from * to end of row. Turn.
Rep 2nd and 3rd rows for Pat until work from beg measures 5½" [14cm], ending with a WS row. Fasten off.
Sew foundation ch and last row tog.

Flower
With Color 2, ch 6. Join with sl st to first ch to form a ring.
1st rnd Ch 1. (1 sc. Ch 2) 6 times in ring. Join with sl st to first sc. 6 ch-2 sps.
2nd rnd Sl st in first ch-2 sp. *3 hdc in same sp as last sl st. Sl st in next ch-2 sp. Rep from * around, ending with 3 hdc in same sp as last sl st. Sl st in same ch-2 sp.
3rd rnd Working behind 2nd rnd, ch 2. (1 sc between next 2 sc. Ch 4) 6 times in ring. Join with sl st to first sc.

4th rnd Sl st in first ch-4 sp. *(1 sc. 4 dc. 1 sc) in same sp as last sl st. Sl st in next ch-4 sp. Rep from * around, ending with (1 sc. 4 dc. 1 sc) in same sp as last sl st. Sl st in same ch-4 sp. Fasten off.

Petal edging Join Color 1 with sl st to first sc of Petal of 4th rnd. Ch 1. 1 sc in same sp as sl st. *1 sc in each of next 4 dc. 1 sc in next sc. Sl st in next sl st. 1 sc in next sc. Rep from * around, ending with 1 sc in each of next 4 dc. 1 sc in next sc. Join with sl st to first sc. Fasten off.

Make 1 Napkin Ring having A as Color 1, B as Color 2.
Make 1 Napkin Ring having B as Color 1, C as Color 2.
Make 1 Napkin Ring having C as Color 1, A as Color 2.

Sew Flower to Napkin Ring to cover seam.

NAPKIN RING DIAGRAM

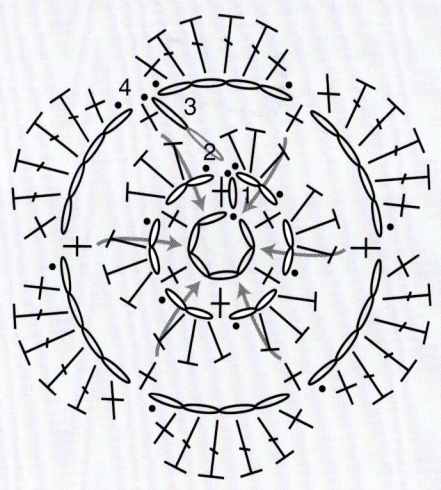

STITCH KEY
⌒ = chain (ch)
• = slip st (sl st)
+ = single crochet (sc)
T = half double crochet (hdc)
† = double crochet (dc)

GROCERY bag

YOU'LL NEED

YARN
Lily® Sugar'n Cream® Confectionary Colors Solids 14oz [400g], 710yds, [649m] and Ombres 12oz [340g], 608yds [556m]

Main Color (MC)
1 ball 27244 (Pistachio)

Contrast (A)
1 ball 33622 (Fruit Cake)

HOOK
Size H (5.0mm) crochet hook *or size to obtain gauge*

MEASUREMENTS
Approx 13" x 7" x 11" [33 x 18 x 28cm].

GAUGE
14 dc and 6 rows = 4" [10cm]. *Remember to check your gauge!*

BAG
Front and back
(make 2 pieces alike)
With MC, ch 41.
1st row (RS) 1 dc in 4th ch from hook (counts as 2 dc). 1 dc in each ch across. 39 dc. Turn.
2nd row Ch 3 (counts as dc). 1 dc in each dc to end of row. Turn.
Rep last row until work from beg measures 13" [33cm]. Fasten off.

Front pocket
With MC, ch 39.
1st row (RS) 1 dc in 4th ch from hook (counts as 2 dc). *(Ch 1. Skip next ch. 1 dc in next ch) 4 times. Ch 1. Skip next ch.** 1 dc in each of next 3 ch. Rep from * once more, then from * to ** once. 1 dc in each of last 2 ch. Turn—37 sts.
2nd row Ch 3 (counts as dc). 1 dc in next dc. *(Ch 1. Skip next ch-1 sp. 1 dc in next dc) 4 times. Ch 1. Skip next ch-1 sp.** 1 dc in each of next 3 dc. Rep from * once more, then from * to ** once. 1 dc in next dc.
1 dc in top of turning ch. Turn.
Rep last row until piece measures 12" [30.5cm]. Fasten off.

Gusset and base
With A, ch 26.
1st row (RS) 1 dc in 4th ch from hook (counts as 2 dc). 1 dc in each ch across. 24 dc. Turn.
2nd row Ch 3 (counts as dc). 1 dc in each dc across, ending with 1 dc in top of turning chain. Turn.
Rep last row until work from beg measures 37" [94cm]. Fasten off.
Place markers on side edges 13" [33cm] and 24" [61cm] down from foundation row.

Handles (make 2)
With A, ch 13.
1st row 1 dc in 4th ch from hook (counts as 2 dc). 1 dc in each of next 9 ch. 11 dc. Turn.
2nd row Ch 3. 1 dc in each dc to end of row. Turn.
Rep last row until Handle measures 20" [51cm]. Do not fasten off.
Fold Handle in half lengthwise and work 1 row of sc through both thicknesses down length of Handle to join sides. Fasten off.

FINISHING
Sew Front Pocket to Front, placing top edge of Pocket 1" [2.5cm] below top edge of Front. To divide pocket into 3 sections, sew down center dc of each 3 dc group. Pin Front and Back pieces to Gusset and Base matching markers to corners. Join MC with sl st to top corner of Front or Back. With WS tog and MC, work 1 rnd of sc around Front and Back to join Gusset and Base. Sew Handles in position.

BOTTLE Bag

YOU'LL NEED

YARN (4)
Lily® Sugar'n Cream® Confectionary Colors Solids 14oz [400g], 710yds [649m] and Ombres 12oz [340g], 608yds [556m]

Main Color (MC)
1 ball 27620 (Custard)

Contrast (A)
1 ball 33622 (Fruit Cake)

HOOK
Size H (5.0mm) crochet hook *or size to obtain gauge*

MEASUREMENTS
Approx 13" x 7" x 11" [33 x 18 x 28cm].

GAUGE
14 dc and 6 rows = 4" [10cm]. *Remember to check your gauge!*

BAG
Front and back (make 2 pieces alike)
With MC, ch 30.
1st row (RS) 1 dc in 4th ch from hook (counts as 2 dc). 1 dc in each ch across—28 dc. Turn.
2nd row Ch 3 (counts as dc). 1 dc in each dc to end of row. Turn.
Rep last row until work from beg measures 12" [30.5cm]. Fasten off.

Inner divider
With MC, ch 16.
1st row (RS) 1 dc in 4th ch from hook (counts as 2 dc). 1 dc in each ch across—14 dc. Turn.
2nd row Ch 3 (counts as dc). 1 dc in each dc to end of row. Turn.
Rep last row until work from beg measures 12" [30.5cm]. Fasten off.

Gusset and base
With A, ch 16.
1st row (RS) 1 dc in 4th ch from hook (counts as 2 dc). 1 dc in each ch across—14 dc. Turn.
2nd row Ch 3 (counts as dc). 1 dc in each dc to end of row. Turn.
Rep last row until work from beg measures 32" [81.5cm]. Fasten off.

Place markers on side edges 12" [30.5cm] and 20" [51cm] down from foundation row.

Handles (make 2)
With A, ch 13.
1st row 1 dc in 4th ch from hook (counts as 2 dc). 1 dc in each of next 9 ch—11 dc. Turn.
2nd row Ch 3. 1 dc in each dc to end of row. Turn.
Rep last row until Handle measures 18" [45.5cm]. Do not fasten off.
Fold Handle in half lengthwise and work 1 row of sc through both thicknesses down length of Handle to join sides. Fasten off.

FINISHING
Pin Front and Back pieces to Gusset and Base matching markers to corners. Join MC with sl st to top corner of Front or Back. With WS tog and MC, work 1 rnd of sc around Front and Back to join Gusset and Base. Sew Inner Divider to inside of Bag at center to create 2 divisions for bottles. Sew Handles in position.

GIFT BAGS
to knit and crochet

YOU'LL NEED

YARN (4)
Lily® Sugar'n Cream® Ombres 2oz [56.7g], 95yds [87m]

Knit Bag
02016 (Tumbleweed) 2 balls

NEEDLES
Size 4.5 mm (U.S. 7) circular knitting needle 16" [40.5cm] long *or size to obtain gauge*

ADDITIONAL
1yd [91cm] of grosgrain ribbon ½" [1cm] wide

Crochet Bag
2 balls 02244 (Landscape)

HOOK
Size G (4.0mm) crochet hook *or size to obtain gauge*

MEASUREMENTS
Knit Bag Approx 7" x 10" [18 x 25.5cm].
Crochet Bag Approx 9" x 11" [23 x 28cm].

GAUGE
Knit Version 20 sts and 26 rows =4" [10cm] in stocking st.
Crochet Version 17 hdc and 10 rows = 4" [10cm].
Remember to check your gauge!

KNIT BAG
Cast on 70 sts. Join in rnd, placing marker on first st.
Knit in rnds until work from beg measures approx 8" [20.5cm].
Next rnd K3. yo. *K5. yo. Rep from * to last 2 sts. K2—84 sts.
Next rnd Knit, inc 4 sts evenly around—88 sts.

Lacy Chevron Edging
1st rnd *K1. yo. K3. Sl1. K2tog. psso. K3. yo. K1. Rep from * to end of rnd.
2nd rnd Knit.
Rep last 2 rnds for 2" [5cm], ending with a 2nd rnd.
Cast off loosely.

FINISHING
Sew bottom seam of bag.
Weave ribbon through eyelet rnd, tie in a bow.

CROCHET BAG
Ch 72. Join with sl st to first ch to form a ring.
1st rnd Ch 2 (does not count as st). 1 hdc in first ch. 1 hdc in each ch to end of rnd. Join with sl st to first hdc—72 hdc.
2nd rnd Ch 2 (does not count as st). Working through back loops only, 1 hdc in each hdc to end of rnd. Join with sl st to first hdc.
Rep 2nd rnd until work from beg measures 8" [20.5cm].
Next rnd (Eyelet rnd) Ch 2 (does not count as st). 1 hdc in each of first 5 hdc. *Ch 1. Skip next hdc.
1 hdc in each of next 5 hdc. Rep from * to last hdc. Ch 1. Skip last hdc. Join with sl st to first hdc.
Next rnd Ch 2 (does not count as st). 1 hdc in each st or ch-1 sp around. Join with sl st to first hdc.
Next rnd Ch 3. Skip first 2 hdc. 1 sc in next hdc. (Ch 5. Skip next 3 hdc. 1 sc in next hdc) 3 times. *(Ch 3. Skip next 2 hdc. 1 sc in next hdc) twice. (Ch 5. Skip next 3 hdc. 1 sc in next hdc) 3 times. Rep from * to last 3 hdc. Ch 3. Skip last 3 hdc. Join with sl st to first ch of ch 3.
Next rnd Ch 3 (counts as dc). 1 dc in same sp as last sl st. Ch 3. Skip first ch-3 sp. 1 sc in next ch-5 sp. 9 dc in next ch-5 sp. 1 sc in next ch-5 sp. *Ch 3. Skip next ch-3 sp. 3 dc in next sc. Ch 3. Skip next ch-3 sp. 1 sc in next ch-5 sp. 9 dc in next ch-5 sp. 1 sc in next ch-5 sp. Rep from * to last ch-3 sp. Ch 3. 1 dc in same sp as first dc. Join with sl st to top of ch 3.
Next rnd Ch 1. 1 sc in each of first 2 dc. Ch 1. Skip next ch-3 sp and sc. (1 dc in next dc. Ch 1) 9 times. *Skip next ch-3 sp. 1 sc in each of next 3 dc. Ch 1. Skip next ch-3 sp and sc. (1 dc in next dc. Ch 1) 9 times. Rep

from * to last ch-3 sp. 1 sc in next dc. Join with sl st to first sc.
Next rnd Ch 1. 1 sc in first sc. *Skip next sc. (1 dc in next dc. Ch 1) 4 times. (1 dc. Ch 1. 1 dc) in next dc. (Ch 1. 1 dc in next dc) 4 times. Skip next sc. 1 sc in next sc. Rep from * to end of rnd. Join with sl st to first sc. Fasten off.

Drawstring (make 1)
Ch 70. Fasten off.

FINISHING
Sew bottom seam.
Thread drawstring through eyelet rnd. Tie ends of drawstring into knot.

TEA TIME placemat

YOU'LL NEED

YARN (4)
Lily® Sugar'n Cream® Confectionary Colors Solids 14oz [400g], 710yds [649m] and Ombres 12oz [340g], 608yds [556m]

Main Color (MC)
1 ball 27505 (Peach Cobbler)

Contrast A
1 ball 33014 (English Toffee)

Note 1 ball each of MC and A makes 4 Placemats

HOOK
Size H (5.0mm) crochet hook *or size to obtain gauge*

MEASUREMENTS
Approx 18" x 13" [45.5 x 33cm].

GAUGE
15 sts and 13 rows = 4" [10cm] in pat.
Remember to check your gauge!

PLACEMAT
With A, ch 70.
1st row (RS) Draw up a loop in 2nd ch from hook. Draw up a loop in next ch. Yoh and draw through all loops on hook. Ch 1. *(Draw up a loop in next ch) twice. Yoh and draw through all loops on hook. Ch 1. Rep from * to last ch. 1 sc in last ch. Turn—69 sts.
2nd row Ch 1. Draw up a loop in first st. Draw up a loop in next ch-1 sp. Yoh and draw through all loops on hook. Ch 1. *Draw up a loop in next st. Draw up a loop in next ch-1 sp. Yoh and draw through all loops on hook. Ch 1. Rep from * to last st. 1 sc in last st. Turn.
Rep last row for pat 3 times more. Break A. Join MC.
Rep 2nd row for pat until work from beg measures approx 11½" [29cm], ending with a WS row. Break MC. Join A. Work 5 rows in pat. Fasten off.

Pocket
With A, ch 12.
1st row (RS) Draw up a loop in 2nd ch from hook. Draw up a loop in next ch. Yoh and draw through all loops on hook. Ch 1. *(Draw up a loop in next ch) twice. Yoh and draw through all loops on hook. Ch 1. Rep from * to last ch. 1 sc in last ch. Turn—11 sts.
2nd row Ch 1. Draw up a loop in first st. Draw up a loop in next ch-1 sp. Yoh and draw through all loops on hook. Ch 1. *Draw up a loop in next st. Draw up a loop in next ch-1 sp. Yoh and draw through all loops on hook. Ch 1. Rep from * to last st. 1 sc in last st. Turn.
Rep 2nd row for pat until work from beg measures approx 5½" [14cm]. Fasten off.

FINISHING
Sew pocket in position as shown in picture.

KITCHEN rug

YOU'LL NEED

YARN
Lily® Sugar'n Cream® Confectionary Colors Ombres 12oz [340g], 608yds [556m]
2 balls 33615 (Maple Sugar)

HOOK
Size K (6.5mm) crochet hook *or size to obtain gauge*

MEASUREMENTS
Approx 20" x 28" [51 x 71cm].

GAUGE
12 sc and 18 rows = 4" [10cm] with 2 strands tog. *Remember to check your gauge!*

RUG
With 2 strands tog, ch 55.
1st row (RS) 1 sc in 2nd ch from hook. 1 sc in each ch to end of ch. Turn—54 sc.
2nd row Ch 1. 1 sc in first sc. *Ch 4. 1 sc in each of next 4 sc. Rep from * to last sc. Ch 4. 1 sc in last sc. Turn.
3rd row Ch 1. 1 sc in first sc. Skip next ch-4 loop. 1 sc in next sc. Pull just skipped ch-4 loop to RS of work - Bobble made. *1 sc in next sc. Ch 4. 1 sc in each of next 2 sc. Bobble. Rep from * to end of row. Turn.
4th row Ch 1. 1 sc in first sc. *Ch 4. 1 sc in each of next 2 sc. Bobble. 1 sc in next sc. Rep from * to last sc. Ch 4. 1 sc in last sc. Turn.
Rep 3rd and 4th rows until work from beg measures approx 27¾" [70.5cm], ending with a 4th row.
Next row (RS) Ch 1. 1 sc in first sc. *Bobble. 1 sc in each of next 3 sc. Rep from * to last sc. Bobble. Fasten off.

KITCHEN baskets

YOU'LL NEED

YARN
Lily® Sugar'n Cream® Confectionary Colors Solids 14oz [400g], 710yds [649m]

Small Baskets
1 ball 27244 (Pistachio)
1 ball 27620 (Custard)

Medium Basket
1 ball 27201 (Cupcake)

HOOK
Size E (3.5mm) crochet hook *or size to obtain gauge*

ADDITIONAL MATERIALS
Fabric stiffener (optional)

MEASUREMENTS
Small Basket Approx 5" [12.5cm] square x 6" [15cm] deep, with top unfolded.
Medium Basket Approx 5" [12.5cm] wide x 10" [25.5cm] long x 6" [15cm] deep, with top unfolded.

GAUGE
18 sc and 20 rows = 4" [10cm]. *Remember to check your gauge!*

NOTE
The instructions are written for small basket. If changes are necessary for medium basket the instructions will be written thus ( ). When only one number is given, it applies to both baskets. For ease in working, circle all numbers pertaining to your basket.

BASKETS
Sides
Ch 28.
1st row (RS) 1 sc in 2nd ch from hook. 1 sc in each ch to end of ch. Turn—27 sc.
2nd row Ch 1. 1 sc in each sc across. Turn. Rep last row until work from beg measures 5" [12.5cm], ending with a WS row.

Small Basket only:
**Next row (RS) (Fold line). Ch 1. Working through back loops only, 1 sc in each sc across. Turn.
Next row Ch 1. Working through both loops, 1 sc in each sc across. Turn. Rep last row until work from last fold line measures 5" [12.5cm], ending with a WS row.**
Rep from ** to ** twice more.
Next row (RS) (Joining row). Place foundation ch behind last row with WS of work tog. Ch 1. Working through both thicknesses, 1 sc in each sc across. Fasten off.

Medium Basket only:
***Next row (RS) (Fold line) Ch 1. Working through back loops only, 1 sc in each sc across. Turn.
Next row Ch 1. Working through both loops, 1 sc in each sc across. Turn. Rep last row until work from last fold line measures 10" [25.5cm], ending with a WS row.
***Next row (RS) (Fold line) Ch 1. Working through back loops only, 1 sc in each sc across. Turn.
Next row Ch 1. Working through both loops, 1 sc in each sc across. Turn. Rep last row until work from last fold line measures 5" [12.5cm], ending with a WS row.
Rep from *** to *** once more.
Next row (RS) (Joining row). Place foundation ch behind last row, working through both thicknesses, 1 sc in each sc across. Fasten off.

Base
Ch 23 (46).
1st row (RS) 1 sc in 2nd ch from hook. 1 sc in each ch to end of ch. Turn—22 (45) sc.
2nd row Ch 1. 1 sc in each sc across. Turn. Rep last row until work from beg measures 5" [12.5cm], ending with a WS row. Do not fasten off.

Join Base to Sides Place lower edge of Sides behind Base with WS of work tog, aligning corners of Base with fold lines and joining row of Sides.
Next rnd (RS) Ch 1. Working through both thicknesses, work 1 rnd of sc around all 4 sides of Base. Join with sl st in first sc. Fasten off.

FINISHING
Optional Apply fabric stiffener to Baskets following manufacturer's directions. Fold top edge to RS as shown in picture.

KITCHEN organizer

YOU'LL NEED

YARN
Lily® Sugar'n Cream® Confectionary Colors Ombres 12oz [340g], 608yds [556m]

2 balls 33307 (Plum Pudding)

HOOK
Size H (5.0mm) crochet hook *or size to obtain gauge*

ADDITIONAL
4 buttons

MEASUREMENTS
Approx 14" x 18" [35.5 x 45.5cm].

GAUGE
15 sts and 13 rows = 4" [10cm] in pat.
Remember to check your gauge!

ORGANIZER
Main piece
Ch 50.
1st row (RS) Draw up a loop in 2nd ch from hook. Draw up a loop in next ch. Yoh and draw through all loops on hook. Ch 1. *(Draw up a loop in next ch) twice. Yoh and draw through all loops on hook. Ch 1. Rep from * to last ch. 1 sc in last ch. Turn—49 sts.
2nd row Ch 1. Draw up a loop in first st. Draw up a loop in next ch-1 sp. Yoh and draw through all loops on hook. Ch 1. *Draw up a loop in next st. Draw up a loop in next ch-1 sp. Yoh and draw through all loops on hook. Ch 1. Rep from * to last st. 1 sc in last st. Turn.

Rep 2nd row for pat until work from beg measures 16" [40.5cm], ending with a WS row.

Divide for tabs: Next row (RS) Ch 1. 1 sc in each of next 7 sts. Turn. Leave rem sts unworked.
***Next row** Ch 1. 1 sc in each sc across. Turn. Rep last row until tab measures 4½" [11.5cm].
Next row Ch 1. 1 sc in each of next 2 sc. Ch 3. Skip next 3 sc. 1 sc in each of last 2 sc. Turn.
Next row Ch 1. 1 sc in each of next 2 sc. 3 sc in next ch-3 sp. 1 sc in each of last 2 sc. Turn.
Next 3 rows Ch 1. 1 sc in each sc across. Turn. Fasten off.*

**With RS of work facing, skip next 7 sts. Join yarn with sl st in next st. Ch 1. 1 sc in same sp as sl st. 1 sc in each of next 6 sts. Turn. Leave rem sts unworked.
Rep from * to * as given above.**
Rep from ** to ** twice more. 4 tabs complete.

Pocket (make 3)
Ch 44.
1st row (RS) Draw up a loop in 2nd ch from hook. Draw up a loop in next ch. Yoh and draw through all loops on hook. Ch 1. *(Draw up a loop in next ch) twice. Yoh and draw through all loops on hook. Ch 1. Rep from * to last ch. 1 sc in last ch. Turn—43 sts.
2nd row Ch 1. Draw up a loop in first st. Draw up a loop in next ch-1 sp. Yoh and draw through all loops on hook. Ch 1. *Draw up a loop in next st. Draw up a loop in next ch-1 sp. Yoh and draw through all loops on hook. Ch 1. Rep from * to last st. 1 sc in last st. Turn.
Rep 2nd row for pat until work from beg measures 3½" [9cm]. Fasten off.

FINISHING
Sew pockets in position. Sew buttons in position to correspond to buttonholes in tabs.

LOOPY dishcloths

YOU'LL NEED

YARN
Lily® Sugar'n Cream®
Solids 2.5oz [70.9g], 120yds [110m]
1 ball 01740 (Hot Pink)
or 01628 (Hot Orange)
or 01712 (Hot Green)

OR Lily® Sugar'n Cream®
Ombres 2oz [56.7g], 95yds [87m]
1 ball 02739 (Over the Rainbow)

HOOK
Size G (4.0mm) crochet hook *or size to obtain gauge*

ADDITIONAL
Cardboard 1" [2.5cm] wide and 9" [23cm] long

MEASUREMENT
Approx 9" [23cm] square.

GAUGE
15 sc and 16 rows = 4" [10cm]. *Remember to check your gauge!*

NOTE
Refer to ABBREVIATIONS (page 48) for Loop.

DISHCLOTH
Ch 35.
1st row (RS) 1 sc in 2nd ch from hook. 1 sc in each ch to end of ch. Turn—34 sc.
2nd row Ch 1. 1 sc in first sc. *Loop. 1 sc in next sc. Rep from * across. Turn.
3rd row Ch 1. *1 sc in next sc. Skip next ch. Rep from * to end of row. Turn.
Rep 2nd and 3rd rows until work from beg measures 9" [23cm], ending with 2nd row. Fasten off.

SPRINGTIME triangle

YOU'LL NEED

YARN
Lily® Sugar'n Cream®
Solids 2.5oz [70.9g], 120yds [110m]
OR
Ombres 2oz [56.7g], 95yds [87m]

Main Color (MC)
1 ball 00001 (White) or 01712 (Hot Green) or 01742 (Hot Blue)

Contrast (A)
1 ball 00010 (Yellow), 00001 (White), 01742 (Hot Blue)
or 02743 (Summer Splash)
or 02744 (Swimming Pool)
or 00165 (Daisy)

HOOK
Size H (5.0mm) crochet hook *or size to obtain gauge*

MEASUREMENT
Approx 12½" [32cm] each side of triangle.

GAUGE
13 sc and 14 rows = 4" [10cm]. *Remember to check your gauge!*

DISHCLOTH
Center Triangle
With MC, ch 21.
1st row (RS) 1 sc in 2nd ch from hook. 1 sc in each ch to end of ch. Turn—20 sc.
2nd row Ch 1. 1 sc in each st to end of row. Turn.
3rd row Ch 1. 1 sc in each sc to last 2 sc. Draw up a loop in each of next 2 sts. Yo and draw through all 3 loops on hook—sc2tog over last 2 sc made. Turn.
4th to 18th rows As 3rd row 15 times. 4 sts at end of last row.
19th row Ch 1. (Sc2tog over next 2 sc) twice. Turn.
20th row Ch 1. Sc2tog over next 2 sts. Fasten off.

First Triangle
1st row With RS of work facing, join A with sl st to corner ch of foundation chain of Center Triangle. Ch 1. Working into rem loops of foundation ch, 1 sc in each ch to end of ch. Turn.
Work 2nd to 20th rows as given for Center Triangle.
Note First Triangle should match Center Triangle.

Second Triangle
1st row With RS of work facing, turn work so First Triangle is to the right. Join A with sl st to top corner of Center Triangle. Ch 1. Work 20 sc evenly across. Turn.
Work 2nd to 20th rows as given for Center Triangle.

Third Triangle
1st row With RS facing, join A with sl st to rem side of Center Triangle. Ch 1. Work 20 sc evenly across. Turn.
Work 2nd to 20th rows as given for Center Triangle.

Border and Loop
With RS of work facing, join MC with sl st to corner of any Triangle and work 1 rnd of sc, having 3 sc in corners. Ch 6 (loop). Join with sl st to first sc. Fasten off.

PRETTY pastels

YOU'LL NEED

YARN
Lily® Sugar'n Cream®
Solids 2.5oz [70.9g], 120yds [110m]
OR Ombres 2oz [56.7g], 95yds [87m]

1 ball 00010 (Yellow) or 00093 (Soft Violet) or 00199 (Pretty Pastels)

or 1 ball Lily® Sugar'n Cream® Twists 2oz [56.7g] 20420 (Rose Twists)

HOOK
Size H (5.0mm) crochet hook *or size to obtain gauge*

MEASUREMENT
Approx 9" [23cm] square.

GAUGE
13 sc and 14 rows = 4" [10cm]. *Remember to check your gauge!*

DISHCLOTH

First Section
Ch 3.
1st row (RS) 1 sc in 2nd ch from hook. 1 sc in last ch. Ch 1. Turn—2 sc.
2nd row Working in back loops only, 1 sc in first sc. 2 sc in last sc. Turn—3 sc.
3rd row Ch 1. Working in back loops only, 2 sc in first sc. 1 sc in each sc to end of row. Turn.
4th row Ch 1. Working in back loops only, 1 sc in each sc to last sc. 2 sc in last sc. Ch 1. Turn.
5th to 18th rows As 3rd and 4th rows 7 times.
19th row As 3rd row. 20 sc. Do not break yarn. Do not turn. Rotate clockwise 90°.

Second Section
1st row (RS) Ch 1. Working along straight edge, 20 sc evenly across. Turn.
2nd row Ch 1. Draw up a loop in back loops of each of first 2 sc. Yoh and draw through all 3 loops on hook – sc2tog over first 2 sc made. 1 sc in back loops only of each sc to end of row. Turn.
3rd row Ch 1. Working in back loops only, 1 sc in each sc to last 2 sc. Sc2tog over last 2 sc. Turn.
4th to 19th rows As 2nd and 3rd rows 8 times.
20th row Ch 1. Sc2tog. Do not break yarn. Do not turn. Rotate clockwise 90°.

Third Section
1st row (RS) Ch 1. Working along straight edge, 20 sc evenly across. Turn. Work 3rd and 2nd rows as given for Second Section, 9 times, then 20th row. Do not turn. Rotate clockwise 90°.

Fourth Section
1st row (RS) Working along straight edge, ch 1. 20 sc evenly across. 1 sc in last sc of 19th row of First Section. 1 sc in next st of 19th row. Turn.
2nd row Skip first 2 sc. Working in back loops only, 1 sc in each sc to last 2 sc. Sc2tog over last 2 sc. Turn.
3rd row Sc2tog over first 2 sc. Working in back loops only, 1 sc in each sc to end of row. 1 sc in each of next 2 sc of 19th row of First Section. Turn.
4th to 19th rows As 2nd and 3rd rows 8 times.
20th row Skip first 2 sc. Sc2tog. Do not break yarn. Do not turn.

Edging
1st rnd Work 20 sc along each side of dishcloth, having 3 sc in corners. Join with sl st to first sc.
2nd rnd Ch 1. *3 sc in next sc. Sl st in next sc. Rep from* around. Join with sl st to first sl st. Fasten off.

PRIMARY colors dishcloth

YOU'LL NEED

YARN (4)
Lily® Sugar'n Cream®
Solids 2.5oz [70.9g], 120yds [110m]

Main Color (MC)
1 ball 00001 (White) or 00010 (Yellow) or 00028 (Delft Blue) or 00095 (Red)

Contrast A
1 ball 00028 (Delft Blue) or 00095 (Red) or 00001 (White) or 00010 (Yellow)

Contrast B
1 ball 00010 (Yellow) or 00001 (White) or 00028 (Delft Blue) or 00095 (Red)

Contrast C
1 ball 00095 (Red) or 00028 (Delft Blue) or 00010 (Yellow) or 00001 (White)

HOOK
Size J (6mm) crochet hook *or size to obtain gauge*

MEASUREMENT
Approx 7" [18cm] in diameter.

GAUGE
11 sc and 12 rows = 4" [10cm]. *Remember to check your gauge!*

DISHCLOTH

First Half
With MC, ch 13.
1st row 1 sc in 2nd ch from hook. 1 sc in each ch to end of ch. Turn—12 sc.
2nd row Ch 1. 1 sc in each of first 2 sc. 2 sc in next sc. 1 sc in next 6 sc. 2 sc in next sc. 1 sc in each of last 2 sc. Turn—14 sc.
3rd row Ch 1. 1 sc in each of first 4 sc. 2 sc in next sc. 1 sc in each of next 5 sc. 2 sc in next sc. 1 sc in each of last 3 sc. Turn. 16 sc.
4th row Ch 1. (1 sc in each of next 6 sc. 2 sc in next sc) twice. 1 sc in each sc to end of row. Turn—18 sc.
5th row Ch 1. 1 sc in first sc. 2 sc in next sc. (1 sc in each of next 6 sc. 2 sc in next sc) twice. 1 sc in each sc to end of row. Turn—21 sc.
6th row Ch 1. 1 sc in first sc. 2 sc in next sc. (1 sc in each of next 5 sc. 2 sc in next sc) 3 times. 1 sc in last sc. Turn—25 sc.
7th row Ch 1. 1 sc in first sc. 2 sc in next sc. (1 sc in each of next 6 sc. 2 sc in next sc) 3 times. 1 sc in each sc to end of row. Turn—29 sc.
8th row Ch 1. 2 sc in first sc. (1 sc in each of next 6 sc. 2 sc in next sc) 4 times. Turn—34 sc.
9th row Ch 1. 2 sc in first sc. (1 sc in each of next 6 sc. 2 sc in next sc) 4 times. 1 sc in each of next 4 sc. 2 sc in last sc. Turn—40 sc.
10th row Ch 1. 2 sc in first sc. 1 sc in each of next 3 sc. (2 sc in next sc. 1 sc in each of next 6 sc) 5 times. 2 sc in last sc. Turn—47 sc. Join A. Break MC.
11th row With A, ch 1. 1 sc in first sc. 2 sc in next sc. (1 sc in each of next 6 sc. 2 sc in next sc) 6 times. 1 sc in each sc to end of row. Turn—54 sc. Join B. Break A.
12th row With B, ch 1. 1 sc in first sc. 2 sc in next sc. (1 sc in each of next 6 sc. 2 sc in next sc) 7 times. 1 sc in each of last 3 sc. Turn—62 sc. Join C. Break B.
13th row With C, ch 1. 1 sc in each sc to end of row. Fasten off.

Second Half
With RS facing, join MC with sl st in rem loop of foundation ch of First Half. Ch 1. Work 12 sc evenly across foundation ch. Turn.
Rep 2nd to 13th rows of First Half.

DISHCLOTH and scrubby

YOU'LL NEED

YARN (4)
Lily® Sugar'n Cream®
2.5oz [70.9g], 120yds [110m]
1 ball 00010 (Yellow) or 01742 (Hot Blue) or 01712 (Hot Green)

or Lily® Sugar'n Cream® Ombres 2oz [56.7g], 95yds [87m]
1 ball (02743 Summer Splash)

HOOK
Size G (4.0mm) crochet hook *or size to obtain gauge*

ADDITIONAL
Copper scrubby

MEASUREMENT
Approx 9" [23cm] square.

GAUGE
15 sc and 16 rows = 4" [10cm]. *Remember to check your gauge!*

DISHCLOTH
Join yarn with sl st at side of Scrubby under 3 loops.

1st rnd Ch 1. 1 sc in same sp as sl st. Work 39 sc more around Scrubby. Join with sl st to first sc—40 sc.

2nd rnd Ch 3 (counts as dc). 1 dc around ch 3. 1 dc in same sp as last sl st. 1 dc around last dc made. *(Skip next sc. 1 dc in next sc—1 dc around last dc made—Wrap st made) 4 times. Skip next sc.** 3 Wrap sts in next sc (corner). Rep from * twice more, then from * to ** once. Wrap st in first sc. Join with sl st to top of ch 3—28 Wrap sts.

3rd rnd Sl st in center of first Wrap st. Ch 3 (counts as dc). 1 dc around ch 3. Wrap st in same sp as last sl st. *(Wrap st in center of next Wrap st) 6 times.** 3 Wrap sts in next Wrap st (corner). Rep from * twice more, then from * to ** once. Wrap st in first Wrap st. Join with sl st to top of ch 3—36 Wrap sts.

4th rnd Sl st in center of first Wrap st. Ch 3 (counts as dc). 1 dc around ch 3. Wrap st in same sp as last sl st. *(Wrap st in center of next Wrap st) 8 times.** 3 Wrap sts in next Wrap st (corner). Rep from * twice more, then from * to ** once. Wrap st in first Wrap st. Join with sl st to top of ch 3—44 Wrap sts.

5th rnd Sl st in center of first Wrap st. Sl st in sp between Wrap sts. Ch 4 (counts as dc and ch 1). 1 dc in same st as last sl st. *(1 dc. Ch 1. 1 dc) in next sp between Wrap sts—V-st made. Rep from * around. Join with sl st to 3rd ch of ch 4.

6th rnd Sl st in first ch-1 sp. Ch 2 (does not count as st). 2 hdc in same sp as last sl st. *2 hdc in next ch-1 sp. Rep from * around. Join with sl st to first hdc. Fasten off.

SPRING dreams dishcloth

YOU'LL NEED

YARN (4)
Main Color (MC)
Lily® Sugar'n Cream®
Ombres 2oz [56.7g], 95yds [87m]
1 ball 02216 (Lavender Ice)
or 02027 (Spring Swirl)
or 02510 (Rosewood)
or 02244 (Landscape)

Contrast A
Lily® Sugar'n Cream®
Solids 2.5oz [70.9g], 120yds [110m]
1 ball 01215 (Robin's Egg Blue)
or 00093 (Soft Violet)
or 00046 (Rose Pink)
or 01011 (Softly Taupe)

HOOK
Size G (4.0mm) crochet hook *or size to obtain gauge*

MEASUREMENT
Approx 9½" [24cm] square.

GAUGE
15 sc and 16 rows = 4" [10cm]. *Remember to check your gauge!*

DISHCLOTH
Center With MC, ch 24.
1st row (RS) 1 dc in 4th ch from hook (counts as 2 dc). 1 dc in each ch to end of chain—22 dc.
2nd row Ch 3 (counts as dc). *Dcfp around post of next dc. Dcbp around post of next dc. Rep from * to last dc. 1 dc in last dc. Turn.
3rd row Ch 3. *Dcbp around post of next st. Dcfp around post of next st. Rep from * to last dc. 1 dc in last dc. Turn.
4th to 11th rows Rep 2nd and 3rd rows 4 times more.
12th row As 2nd row. Fasten off.

Edging
1st rnd With RS of work facing, join A with sl st in any corner. (Ch 3. 1 dc. Ch 2. 2 dc) all in same sp as sl st. *1 dc in each dc across to next corner.** (2 dc. Ch 2. 2 dc) in corner. Rep from * twice more, then from * to ** once. Join with sl st in top of ch 3.
2nd rnd Sl st in next dc and ch-2 sp. (Ch 3. 1 dc. Ch 2. 2 dc) all in same sp as last sl st. *1 dc in each dc to next ch-2 sp.** (2 dc. Ch 2. 2 dc) in next ch-2 sp. Rep from* twice more, then from * to ** once. Join with sl st in top of ch 3. Fasten off.
3rd rnd Join MC with sl st in any dc. Ch 1. Work 1 sc in each dc and 3 sc in each ch-2 sp around. Join with sl st in first sc.
4th rnd Ch 1. 1 sc in each sc around, working 3 sc in each corner sc. Join with sl st to first sc. Fasten off.
5th rnd Join A with sl st in any sc. Ch 3 (counts as dc). *1 dc in each sc to next corner sc. (2 dc. Ch 2. 2 dc) in corner sc. Rep from * 3 times more. 1 dc in each dc to end of rnd. Join with sl st in top of ch 3. Fasten off.

WITH a twist

YOU'LL NEED

YARN (4)
Lily® Sugar'n Cream®
Solids 2.5oz [70.9g], 120yds [110m]

2-COLOR VERSION
Main Color (MC)
1 ball 00010 (Yellow)
or 00028 (Deft Blue)
Contrast (A)
1 ball or 00001 (White)

1-COLOR VERSION
Main Color (MC)
1 ball or 00001 (White)

HOOK
Size I (5.5mm) crochet hook *or size to obtain gauge*

MEASUREMENT
Approx 7½" [19cm] square.

GAUGE
12 sc and 13 rows = 4" [10cm]. *Remember to check your gauge!*

2-COLOR VERSION

Square
With MC, ch 4. Join in rnd with sl st to form a ring.
1st rnd Ch 1. [(1 sc. Ch 2) 3 times. 1 sc] in ring. Ch 1. 1 hdc in first sc.
2nd rnd Ch 1. (1 sc. Ch 2. 1 sc) in top of hdc. [(Ch 2. 1 sc) twice] 3 times in next ch-2 sp. Ch 1. 1 hdc in first sc.
3rd rnd Ch 1. 1 sc in top of hdc. *[(Ch 2. 1 sc) twice] in next ch-2 sp for corner.** Ch 2. 1 sc in next ch-2 sp. Rep from * twice more, then from * to ** once. Ch 1. 1 hdc in first sc.
4th rnd Ch 1. 1 sc in top of hdc. *Ch 2. 1 sc in next ch-2 sp. [(Ch 2. 1 sc) twice] in next ch-2 sp for corner.** Ch 2. 1 sc in next ch-2 sp. Rep from * twice more, then from * to ** once. Ch 1. 1 hdc in first sc.
5th rnd Ch 1. 1 sc in top of hdc. *(Ch 2. 1 sc in next ch-2 sp) twice. [(Ch 2. 1 sc) twice] in next ch-2 sp for corner.** Ch 2. 1 sc in next ch-2 sp. Rep from * twice more, then from * to ** once. Ch 1. 1 hdc in first sc.
6th rnd Ch 1. 1 sc in top of hdc. *(Ch 2. 1 sc in next ch-2 sp) 3 times. [(Ch 2. 1 sc) twice] in next ch-2 sp for corner.** Ch 2. 1 sc in next ch-2 sp. Rep from * twice more, then from * to ** once. Ch 1. 1 hdc in first sc.
7th rnd Ch 1. 1 sc in top of hdc. *(Ch 2. 1 sc in next ch-2 sp) 4 times. [(Ch 2. 1 sc) twice] in next ch-2 sp for corner.** Ch 2. 1 sc in next ch-2 sp. Rep from * twice more, then from * to ** once. Ch 1. 1 hdc in first sc.
8th rnd Ch 1. 2 sc in top of hdc. *(2 sc in next ch-2 sp) 5 times. 3 sc in next corner ch-2 sp.** 2 sc in next ch-2 sp. Rep from * twice more, then from * to ** once. Join with sl st to first sc. Fasten off.

Bow
With RS of work facing, join A with sl st to front loop of any corner sc.
1st row Ch 1. Working in front loops only, 1 sc in same sp as sl st. 1 sc in each of next 15 sc. Turn. Leave rem sts unworked—16 sc.
2nd row Ch 1. Draw up a loop in each of first 2 sc. Yoh and draw through all loops on hook – Sc2tog over first 2 sc made. 1 sc in each of next 12 sc. Sc2tog over last 2 sc. Turn—14 sts.
3rd and 4th rows Ch 1. 1 sc in each st to end of row. Turn.
5th row Ch 1. Sc2tog over first 2 sc. 1 sc in each of next 10 sc. Sc2tog over last 2 sc. Turn—12 sts.
6th row Ch 1. Sc2tog over first 2 sts. 1 sc in each of next 8 sc. Sc2tog over last 2 sts. Turn—10 sts.
7th row Ch 1. Sc2tog over first 2 sts. 1 sc in each of next 6 sc. Sc2tog over last 2 sts. Turn—8 sts.
8th row Ch 1. Sc2tog over first 2 sts. 1 sc in each of next 4 sc. Sc2tog over last 2 sts. Turn—6 sts.
9th row Ch 1. Sc2tog over first 2 sts. 1 sc in each of next 2 sc. Sc2tog over last 2 sts. Turn—4 sts.
10th to 12th rows Ch 1. 1 sc in each st to end of row. Turn.
13th row Ch 1. 1 sc in first sc. 2 sc in each of next 2 sc. 1 sc in last sc. Turn—6 sc.
14th row Ch 1. 1 sc in first sc. 2 sc in next sc. 1 sc in each sc to last 2 sc. 2 sc in next sc. 1 sc in last sc. Turn.

Rep last row 3 times more—14 sc.
Next 3 rows Ch 1. 1 sc in each sc to end of row. Turn.
Next row As 14th row—16 sc. Do not turn.

Joining Bow to Square
Twist Bow in center.
Align last row of Bow and opposite side of Square. Working through both loops of Bow and front loops of Square, proceed as follows:
Next row Sl st in each st across row. Fasten off.
Note Center twist of bow will flatten once you are done joining.

Border
9th rnd With RS of work facing, join MC with sl st to corner sc of Square. Ch 1. 3 sc in same sp as sl st. *1 sc in each sc to next corner sc.** 3 sc in corner sc. Rep from * twice more, working in back loops along joining rows, then from * to ** once. Join with sl st to first sc.
10th rnd Ch 2 (does not count as st). Work 1 hdc in each sc around, having 3 hdc in corners. Join with sl st to first hdc.
11th rnd As 10th rnd. Join A.
12th rnd With A, ch 1. 1 sc in same sp as sl st. **Ch 2. Skip next hdc. 1 sc in next hdc. Rep from * around. Ch 2. Join with sl st to first sc. Fasten off.

1-COLOR VERSION
Work as given for 2-Color Version, omitting all references to color changes.

LACY dishcloth

YOU'LL NEED

YARN (4)
Lily® Sugar'n Cream® Confectionary
Colors Solids 14oz [400g], 710yds,
[649m] and
Ombres 12oz [340g], 608yds [556m]

1 ball each
A 27244 (Pistachio)
B 27620 (Custard)
C 33615 (Maple Sugar)
D 27427 (Cinnamon)

Note Solids: 1 ball makes 11 of dishcloth A or B or D.
Ombre: 1 ball makes 12 of dishcloth C

HOOK
Size H (5.0mm) crochet hook *or size to obtain gauge*

MEASUREMENTS

Dishcloth A Approx 9½" [24cm] square.
Dishcloth B Approx 9½ " [24cm] square.
Dishcloth C Approx 9" [23cm] square.
Dishcloth D Approx 10½" [26.5cm] square.

GAUGE

14 sc and 15 rows = 4" [10cm]. *Remember to check your gauge!*

DISHCLOTH A

Ch 34.
1st row (WS) 1 sc in 2nd ch from hook. 1 sc in each ch across. Turn. 33 sc.
2nd row Ch 3 (counts as dc). 1 dc in next sc. *Skip next 2 sc. 1 dc in next sc. Ch 3 (counts as dc). Work 3 dc down side of dc just worked to form block. Skip next 2 sc. 1 dc in each of next 3 sc. Rep from * across, ending last rep with 1 dc in each of last 2 sc. Turn.
3rd row Ch 3 (counts as dc). 1 dc in next dc. *Ch 2. 1 sc in top of ch 3 at corner of next block. Ch 2. 1 dc in each of next 3 dc. Rep from * across, ending with 1 dc in last dc. 1 dc in top of ch 3. Turn.
4th row Ch 3 (counts as dc). 1 dc in next dc. *1 dc in next sc. Ch 3 (counts as dc). Work 3 dc down side of dc just worked to form block. 1 dc in each of next 3 dc. Rep from * across, ending with 1 dc in last dc. 1 dc in top of ch 3. Turn.
Rep 3rd and 4th rows 5 times more, then 3rd row once.
Next row Ch 1. 1 sc in each of next 2 dc. *2 sc in next ch-2 sp. 1 sc in next sc. 2 sc in next ch-2 sp. 1 sc in each of next 3 dc. Rep from * across, ending with 1 sc in last dc. 1 sc in top of turning ch. Fasten off.

DISHCLOTH B

Ch 31.
1st row (WS) 1 sc in 2nd ch from hook. 1 sc in next ch. *Ch 3. Skip next 3 ch. 1 sc in each of next 3 ch. Rep from * to last 5 ch. Ch 3. Skip next 3 ch. 1 sc in each of last 2 ch. Turn.
2nd row Ch 1. 1 sc in first sc. *5 dc in next ch-3 sp. Skip next sc. 1 sc in next sc. Rep from * to end of row. Turn.
3rd row Ch 1. 1 sc in first sc. Ch 2. *1 sc in 2nd, 3rd and 4th dc of next 5 dc group. Ch 3. Rep from *, ending last rep with ch 2. 1 sc in last sc. Turn.
4th row Ch 3 (counts as dc). 2 dc in next ch-2 sp. Skip next sc. 1 sc in next sc. *5 dc in next ch-3 sp. Skip next sc. 1 sc in next sc. Rep from * to last ch-2 sp. 2 dc in last ch-2 sp. 1 dc in last sc. Turn.
5th row Ch 1. 1 sc in each of first 2 dc. *Ch 3. 1 sc in 2nd, 3rd and 4th dc of next 5 dc group. Rep from *, ending with ch 3. Skip next dc. 1 sc in next dc. 1 sc in top of ch 3. Turn.
Rep 2nd to 5th rows 3 times more. Fasten off.

DISHCLOTH C

Ch 6. Join with sl st in first ch to form a ring.
1st rnd Ch 3 (counts as dc). 15 dc in ring. Join with sl st to top of ch 3. 16 dc.
2nd rnd Ch 3 (counts as dc). 2 dc in same sp as last sl st. Ch 2. Skip next dc. 1 dc in next dc. Ch 2. Skip next dc. *3 dc in next dc. Ch 2. Skip next dc. 1 dc in next dc. Ch 2. Skip next dc. Rep from * twice more. Join with sl st to top of ch 3.
3rd rnd Ch 3 (counts as dc). 5 dc in next dc. *1 dc in next dc. (Ch 2. 1 dc in next dc) twice. 5 dc in next dc. Rep from * twice more. (1 dc in next dc. Ch 2) twice. Join with sl st to top of ch 3.
4th rnd Ch 3 (counts as dc). 1 dc in each of next 2 dc. 5 dc in next dc. *1 dc in each of next 3 dc. Ch 2. 1 dc in next dc. Ch 2.** 1 dc in each of next 3 dc. 5 dc in next dc. Rep from * twice more, then from * to ** once. Join with sl st to top of ch 3.
5th rnd Ch 3 (counts as dc). 1 dc in each of next 4 dc. 5 dc in next dc. *1 dc in each of next 5 dc. Ch 2. 1 dc in next dc. Ch 2.** 1 dc in each of next 5 dc. 5 dc in next dc. Rep from * twice more, then from * to ** once. Join with sl st to top of ch 3.
6th rnd Ch 3 (counts as dc). 1 dc in each of next 6 dc. 5 dc in next dc. *1 dc in each of

next 7 dc. Ch 2. 1 dc in next dc. Ch 2.** 1 dc in each of next 7 dc. 5 dc in next dc. Rep from * twice more, then from * to ** once. Join with sl st to top of ch 3.
7th rnd Ch 3 (counts as dc). 1 dc in each dc and 2 dc in each ch-2 sp around, working (2 dc. Ch 1. 2 dc) in center dc of each 5 dc corner group. Join with sl st to top of ch 3. Fasten off.

DISHCLOTH D

Ch 6. Join with sl st in first ch to form a ring.
1st rnd Ch 3 (counts as dc). 4 dc in ring. Drop loop from hook and insert in top of first dc of 4 dc group. Pull dropped loop through – Popcorn made. (Ch 5. Popcorn) 3 times in ring. Ch 5. Join with sl st to top of ch 3.
2nd rnd Ch 3 (counts as dc). *(2 dc. Ch 2. Popcorn. Ch 2. 2 dc) in next ch-5 sp.** 1 dc in next Popcorn. Rep from * twice more, then from * to ** once. Join with sl st to top of ch 3.
3rd rnd Ch 3. 1 dc in each of next 2 dc. *2 dc in next ch-2 sp. Ch 2. Popcorn in next Popcorn. Ch 2. 2 dc in next ch-2 sp.** 1 dc in each of next 5 dc. Rep from * twice more, then from * to ** once. 1 dc in each of next 2 dc. Join with sl st to top of ch 3.
4th rnd Ch 3. 1 dc in each of next 4 dc. *2 dc in next ch-2 sp. Ch 2. Popcorn in next Popcorn. Ch 2. 2 dc in next ch-2 sp.** 1 dc in each of next 9 dc. Rep from * twice more, then from * to ** once. 1 dc in each of next 4 dc. Join with sl st to top of ch 3.
5th rnd Ch 3. 1 dc in each of next 6 dc. *2 dc in next ch-2 sp. Ch 2. Popcorn in next Popcorn. Ch 2. 2 dc in next ch-2 sp.** 1 dc in each of next 13 dc. Rep from * twice more, then from * to ** once. 1 dc in each of next 6 dc. Join with sl st to top of ch 3.
6th rnd Ch 3. 1 dc in each of next 8 dc. *2 dc in next ch-2 sp. Ch 2. Popcorn in next Popcorn. Ch 2. 2 dc in next ch-2 sp.** 1 dc in each of next 17 dc. Rep from * twice more, then from * to ** once. 1 dc in each of next 8 dc. Join with sl st to top of ch 3.
7th rnd Ch 3. 1 dc in each of next 10 dc. *2 dc in next ch-2 sp. Ch 2. Popcorn in next Popcorn. Ch 2. 2 dc in next ch-2 sp.** 1 dc in each of next 21 dc. Rep from * twice more, then from * to ** once. 1 dc in each of next 10 dc. Join with sl st to top of ch 3. Fasten off.

SPA face cloth

YOU'LL NEED

YARN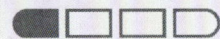

Lily® Sugar 'n Cream Twists
2oz [56.7g], 95yds [87m]

2 balls 20116 (Country Twists) or 20117 (Denim Twists) or 20244 (Green Twists)

HOOK
Size H (5.0mm) crochet hook *or size to obtain gauge*

MEASUREMENTS
Approx 13" [33cm] square.

GAUGE
16 sts and 16 rows = 4" [10cm] in pat.
Remember to check your gauge!

FACE CLOTH
Ch 49.

1st row (RS) 1 sc in 2nd ch from hook. *Ch 1. Skip next ch. 1 sc in next ch. Rep from * to last ch. 1 sc in last ch. Turn—48 sts.

2nd row Ch 1. 1 sc in first sc. *Ch 1. Skip next sc. 1 sc in next ch-1 sp. Rep from * to last sc. 1 sc in last sc. Turn.

Rep last row until work from beg measures 12" [30.5cm], ending with a WS row.

Edging

1st rnd Ch 1. Work 48 sc across top of Face cloth. Do not turn. Work 48 sc evenly along each of rem 3 edges of Face cloth, having 3 sc in corners. Join with sl st in first sc—192 sc.

2nd rnd Ch 1. Skip first sc. *5 dc in next sc. Skip next sc. 1 sc in next sc. Rep from * around. Join with sl st in top of first sc. Fasten off.

PASTEL washcloths

YOU'LL NEED

YARN 4
Lily® Sugar 'n Cream Twists
2oz [56.7g], 95yds [87m]

1 ball
20009 (Taupe Twists)
or 20010 (Natural Twists)
or 20116 (Country Twists)
or 20117 (Denim Twists)
or 20244 (Green Twists)
or 20315 (Summer Twists)
or 20410 (Rose Twists)

HOOK
Size H (5.0mm) crochet hook *or size to obtain gauge*

MEASUREMENT
Approx 10" [25.5cm] square.

GAUGE
12 sc and 13 rows = 4" [10cm]. *Remember to check your gauge!*

WASHCLOTH
Ch 31.
1st row (RS) 1 sc in 2nd ch from hook. 1 sc in each ch to end of ch—30 sc. Turn.
2nd row Ch 1. 1 sc in each sc to end of row. Turn. Rep last row until work from beg measures 10" [25.5cm], ending with a WS row. Fasten off.

SPA set

YOU'LL NEED

YARN (4)
Lily® Sugar 'n Cream Twists
2oz [56.7g], 95yds [87m]

Towel 3 balls/**Bath Mitt** 1 ball
20010 (Natural Twists)

NEEDLES
Size 8 (5mm) and 9 (5.5mm) knitting needles *or size to obtain gauge*

ADDITIONAL
Fabric stiffener

MEASUREMENTS
Towel Approx 15" x 24" [38 x 61cm].
Bath Mitt One size to fit average lady's hand.

GAUGE
Towel 18 sts and 24 rows = 4" [10cm] with smaller needles in stocking st.
Bath Mitt 15 sts and 28 rows = 4" [10cm] with larger needles in garter st.
Remember to check your gauge!

TOWEL
With smaller needles, cast on 61 sts.
Knit 3 rows (garter st), noting first row is WS.
1st row (RS) K2. *K3. P3. Rep from * to last 5 sts. K5.
2nd row K2. *P3. K3. Rep from * to last 5 sts. P3. K2.
3rd and 4th rows As 1st and 2nd rows once more.
5th row K2. *P3. K3. Rep from * to last 5 sts. P3. K2.
6th row K2. *K3. P3. Rep from * to last 5 sts. K5.
7th and 8th rows As 5th and 6th rows once more.
Last 8 rows form pat.
Cont in pat until work from beg measures approx 23½" [59.5cm], ending with 4th row of pat.
Next 3 rows Knit. Cast off knitwise (WS).

BATH MITT
With larger needles, cast on 36 sts.
Knit 9 rows (garter st), noting that first row is WS.

Shape thumb
1st row (RS) K16. Inc 1 st in each of next 2 sts. Knit to end of row.
2nd to 4th rows Knit.
5th row K16. Inc 1 st in next st. K2. Inc 1 st in next st. Knit to end of row.
6th to 8th rows Knit.
9th row K16. Inc 1 st in next st. K4. Inc 1 st in next st. Knit to end of row.
10th to 12th rows Knit.
Cont in this manner, inc 2 sts on next and every following 4th row to 48 sts.
Next row (WS) Knit.

Shape thumb
Next row K29. Turn. Cast on 1 st. K14 (including st on needle after cast on). Turn. Leave rem sts unworked—14 sts.
Knit 10 rows.
Next row (RS) *K1. K2tog. Rep from * to last 2 sts. K2tog—10 sts.
Next row Knit.
Next row (K2tog) 5 times—5 sts. Break yarn, leaving a long end. Thread end through rem sts and draw up tightly. Fasten securely. Sew thumb seam.
With RS of Mitt facing, join yarn to rem sts: pick up and knit 2 sts at base of thumb. Knit to end of row—37 sts.
Next row Knit, working 2tog over 2 sts picked up at base of thumb—36 sts.
Knit 16 rows.

Shape top
1st row (RS) (K1. Sl1. K1. psso. K12. K2tog. K1) twice—32 sts.
2nd and alt rows Knit.
3rd row (K1. Sl1. K1. psso. K10. K2tog. K1) twice—28 sts.
5th row (K1. Sl1. K1. psso. K8. K2tog. K1) twice—24 sts.
7th row (K1. Sl1. K1. psso. K6. K2tog. K1) twice—20 sts.
9th row (K1. Sl1. K1. psso. K4. K2tog. K1) twice—16 sts.
11th row (K2tog) 8 times. 8 sts. Break yarn leaving a long end. Thread end through rem sts and draw up tightly. Fasten securely. Sew side seam.

Hanging Loop With larger needles, cast on 15 sts. Cast off. Sew ends of loop to edge of Mitt at side seam.

SPA Bag

YOU'LL NEED

YARN (4)
Lily® Sugar 'n Cream Twists
2oz [56.7g], 95yds [87m]
5 balls 20009 (Taupe Twists)

HOOK
Size F (3.75mm) crochet hook *or size to obtain gauge*

ADDITIONAL
4 Grommets with openings approx ½" [1cm]

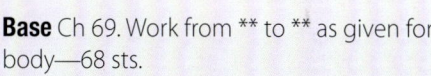

MEASUREMENTS
Approx 15" [38cm] wide x 12" [30.5cm] high, excluding handles.

GAUGE
18 sts and 12 rows = 4" [10cm] in pat. *Remember to check your gauge!*

NOTE
Turning ch 2 at beg of row does not count as hdc.

BAG
Body of bag
Ch 56.
Foundation row (RS) 1 sc in 2nd ch from hook. 1 sc in each ch to end of ch. Turn—55 sc.
1st row Ch 2. Working through both loops, 1 hdc in each st across. Turn.
2nd row Ch 2. Working through back loops only, 1 hdc in each st across. Turn.**
Rep last 2 rows for pat until work from beg measures 4" [10cm], ending with a WS row.
Next row (RS) (Eyelet row) Ch 2. Working through back loops only, 1 hdc in each of first 11 hdc. Ch 1. Skip next hdc. Working through back loops only, 1 hdc in each hdc to end of row. Turn. 55 sts.
Cont in pat until work from beg measures 11" [28cm], ending with a WS row. Place marker at end of last row.
Work Eyelet row once more.
Cont in pat until work from marked row measures 10" [25.5cm], ending with a WS row. Place marker at end of last row.
Work Eyelet row once more.
Cont in pat until work from last marked row measures 7" [18cm].
Work Eyelet row once more. Place marker at end of last row.
Cont in pat until work from last marked row measures 6" [15cm], ending with a WS row. Remove markers.
Measuring from foundation ch, place markers on left edge of Body at beg of 2nd row of pat at approx 15", 17" and 32" [38, 43 and 81.5cm].

Join Body With WS tog, place foundation ch behind last row. Working through both thicknesses, sl st in each st across. Fasten off.

Base Ch 69. Work from ** to ** as given for body—68 sts.
Rep last 2 rows until work from beg measures 2" [5cm], ending with a WS row. Do not fasten off.

Join Base
With WS tog, place marked edge of Body behind Base, aligning markers and joining row with corners of Base. Ch 1. Working through both thicknesses, work 1 rnd of sc evenly around Base. Join with sl st to first sc. Fasten off.

HANDLES (make 2)
Ch 96.
1st row (RS) 1 hdc in 3rd ch from hook. 1 hdc in each ch across. Turn—94 hdc.
2nd row Ch 2. 1 hdc in each hdc across. Turn.
Rep last row once more. Fasten off.
Apply grommets in eyelets following manufacturer's directions. Knot one end of Handle. Thread other end of Handle through both grommets on one side of Bag and knot end as shown in picture. Rep for other Handle.

DRAWER organizer

YOU'LL NEED

YARN
Lily® Sugar 'n Cream Twists
2oz [56.7g], 95yds [87m]

7 balls 20010 (Natural Twists)

HOOKS
Size D (3.25mm) crochet hook *or size to obtain gauge*

ADDITIONAL
Fabric stiffener (optional)

MEASUREMENTS
Approx 10" [25.5cm] square x 6" [15cm] deep with top unfolded.

GAUGE
9 sc and 20 rows = 4" [10cm]. *Remember to check your gauge!*

ORGANIZER
Sides
Ch 31.
1st row (RS) 1 sc in 2nd ch from hook. 1 sc in each ch to end of ch. Turn—30 sc.
2nd row Ch 1. 1 sc in each sc across. Turn. Rep last row until work from beg measures 10" [25.5cm], ending with a WS row.
**Next row (RS) (Fold line) Ch 1. Working through back loops only, 1 sc in each sc across. Turn.
Next row Ch 1. Working through both loops, 1 sc in each sc across. Turn. Rep last row until work from fold line measures 10" [25.5cm], ending with a WS row.**
Rep from ** to ** twice more.
Next row (RS) (Joining row) Place foundation ch behind last row with WS of work tog. Ch 1. Working through both thicknesses, 1 sc in each sc across. Fasten off.

Base
Ch 50.
1st row (RS) 1 sc in 2nd ch from hook. 1 sc in each ch to end of ch. Turn—49 sc.
2nd row Ch 1. 1 sc in each sc across. Turn. Rep last row until work from beg measures 10" [25.5cm], ending with a WS row. Do not fasten off.
Join Base to Sides Place lower edge of Sides behind Base with WS of work tog, aligning corners of Base with fold lines and joining row of Sides. Ch 1. Working through both thicknesses, work 1 rnd of sc around all 4 sides of Base. Join with sl st in first sc. Fasten off.

Divider
First Piece
**Ch 25.
1st row (RS) 1 sc in 2nd ch from hook. 1 sc in each ch to end of ch. Turn—24 sc.
2nd row Ch 1. 1 sc in each sc across. Turn. Rep last row until work from beg measures 5" [12.5cm], ending with a WS row.**
Next row (RS) (Center row) Ch 1. Working through back loops only, 1 sc in each sc across. Turn.
Next row Ch 1. Working through both loops, 1 sc in each sc across. Turn. Rep last row until work from beg measures 10" [25.5cm], ending with a WS row. Fasten off.

Second Piece
Work from ** to ** as given for First Piece.
Next row (Joining row) Ch 1. With RS of both pieces facing, place First Piece behind Second Piece having rem loops of Center row behind last row of Second Piece. *Working through back loops only of Second Piece and rem loops of First Piece, 1 sc in next sc. Rep from * to end of row. Turn.
Next row Ch 1. Working through both loops of Second Piece only, 1 sc in each sc across. Turn.
Rep last row until Second Piece from beg measures 10" [25.5cm], ending with a WS row. Fasten off.

FINISHING
Optional Apply Fabric Stiffener to Organizer and Divider following manufacturer's directions.
Fold top edge of Organizer to RS as shown in picture.
Fold Divider as shown in picture and place inside Organizer. Sew edges of Divider in position to sides of Organizer.

BATH boxes

YOU'LL NEED

YARN
Lily® Sugar 'n Cream Twists
2oz [56.7g], 95yds [87m]

20010 (Natural Twists)
Small - 3 balls
Large - 4 balls

HOOKS
Size D (3.25mm) crochet hook *or size to obtain gauge*

ADDITIONAL
Fabric stiffener (optional)

MEASUREMENTS
Small Approx 5" [12.5cm] square x 6" [15cm] deep with top unfolded.
Large Approx 5" [12.5cm] wide x 10" [25.5cm] long x 6" [15cm] deep with top unfolded.

GAUGE
19 sc and 20 rows = 4" [10cm]. *Remember to check your gauge!*

NOTE
The instructions are written for smaller size. If changes are necessary for larger size the instructions will be written thus ( ). When only one number is given, it applies to both sizes. For ease in working, circle all numbers pertaining to your size.

BOXES
Sides
Small Box only Ch 31.
1st row (RS) 1 sc in 2nd ch from hook. 1 sc in each ch to end of ch. Turn—30 sc.
2nd row Ch 1. 1 sc in each sc across. Turn. Rep last row until work from beg measures 5" [12.5cm], ending with a WS row.
**Next row (RS) (Fold line) Ch 1. Working through back loops only, 1 sc in each sc across. Turn.
Next row Ch 1. Working through both loops, 1 sc in each sc across. Turn. Rep last row until work from fold line measures 5" [12.5cm], ending with a WS row.******
Rep from ** to ** twice more.
Next row (RS) (Joining row) Place foundation ch behind last row with WS of work tog. Ch 1. Working through both thicknesses, 1 sc in each sc across. Fasten off.

Large Box only Ch 31.
1st row (RS) 1 sc in 2nd ch from hook. 1 sc in each ch to end of ch. Turn—30 sc.
2nd row Ch 1. 1 sc in each sc across. Turn.
**Rep last row until work from beg measures 5" [12.5cm], ending with a WS row.
Next row (RS) (Fold line) Ch 1. Working through back loops only, 1 sc in each sc across. Turn.
Next row Ch 1. Working through both loops, 1 sc in each sc across. Turn. Rep last row until work from fold line measures 10" [25.5cm], ending with a WS row.**
Next row (RS) (Fold line) Ch 1. Working through back loops only, 1 sc in each sc across. Turn.
Next row Ch 1. Working through both loops, 1 sc in each sc across. Turn.
Rep from ** to ** once more.
Next row (RS) (Joining row) Place foundation ch behind last row with WS of work tog. Ch 1. Working through both thicknesses, 1 sc in each sc across. Fasten off.

Base
Ch 26 (50).
1st row (RS) 1 sc in 2nd ch from hook. 1 sc in each ch to end of ch. Turn—25 (49) sc.
2nd row Ch 1. 1 sc in each sc across. Turn. Rep last row until work from beg measures 5" [12.5cm], ending with a WS row. Do not fasten off.
Join Base to Sides Place lower edge of Sides behind Base with WS of work tog, aligning corners of Base with fold lines and joining row of Sides. Ch 1. Working through both thicknesses, work 1 rnd of sc around all 4 sides of Base. Join with sl st in first sc. Fasten off.

FINISHING
Optional Apply Fabric Stiffener to Boxes following manufacturer's directions. Fold top edge of Boxes to RS as shown in picture.

TISSUE box cover

YOU'LL NEED

YARN (4)
Lily® Sugar 'n Cream Twists
2oz [56.7g], 95yds [87m]

Small Cover - 1 ball
20116 (Country Twists)

Large Cover - 1 ball
20117 (Denim Twists)

HOOK
Size E (3.5mm) crochet hook *or size to obtain gauge*

MEASUREMENTS
Small To fit tissue box 4½" [11.5cm] square x 5" [12.5cm] high.
Large To fit tissue box 4½" [11.5cm] wide x 9" [23cm] long x 3" [7.5cm] high.

GAUGE
16 hdc and 12 rows = 4" [10cm]. *Remember to check your gauge!*

SMALL BOX COVER
Main Section
**Ch 18.
1st row (RS) 1 sc in 2nd ch from hook. 1 sc in each ch to end of ch. Turn—17 sc.
2nd row Ch 2. 1 hdc in each sc to end of row. Turn.
3rd row Ch 2. 1 hdc in each hdc to end of row. Turn.**
Rep last row 12 times more. Place marker at each end of last row.
Rep last row 6 times more.

Divide for opening
Next row (WS) Ch 2. 1 hdc in each of next 4 hdc. Ch 9. Skip next 9 hdc. 1 hdc in each hdc to end of row. Turn.
Next row Ch 2. 1 hdc in each of next 4 hdc. 1 hdc in each of next 9 ch. 1 hdc in each hdc to end of row. Turn—17 hdc.
Next row Ch 2. 1 hdc in each hdc to end of row. Turn.
Rep last row 4 times more. Place marker at each end of last row.
Rep last row 14 times more.
Next row Ch 1. 1 sc in each hdc to end of row. Fasten off.

Sides (make 2)
Work from ** to ** as given for Main Section.
Work a further 10 rows even. Fasten off.

Joining
With WS of pieces facing each other, pin Sides to Main Section having last row of Side between markers on Main Section. With RS of Main Section facing, join yarn with sl st to lower edge. Ch 1. Work 15 sc through both thicknesses up side, 17 sc across top and 15 sc down other side. Do not turn.
Next row Ch 1. Working from left to right, intead of from right to left as usual, work 1 reverse sc in each sc to end of row. Fasten off.
Rep for other side.

LARGE BOX COVER
Main Section
Ch 36.
1st row (RS) 1 sc in 2nd ch from hook. 1 sc in each ch to end of ch. Turn—35 sc.
2nd row Ch 2. 1 hdc in each sc to end of row.
3rd row Ch 2. 1 hdc in each hdc to end of row.
Rep last row 6 times more. Place marker at each end of last row.

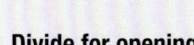

Divide for opening
Next row (WS) Ch 2. 1 hdc in each of next 7 hdc. Ch 21. Skip next 21 hdc. 1 hdc in each hdc to end of row. Turn.
Next row Ch 2. 1 hdc in each of next 7 hdc. 1 hdc in each of next 21 ch. 1 hdc in each hdc to end of row. Turn—35 hdc.
Next row Ch 2. 1 hdc in each hdc to end of row.
Rep last row 3 times more. Place marker at each end of last row.
Rep last row 8 times more.
Next row Ch 1. 1 sc in each hdc to end of row. Fasten off.

Sides (make 2)
Work from ** to ** as given for Main Section of Small Box Cover.
Work a further 6 rows even. Fasten off.

Joining
With WS of pieces facing each other, pin Sides to Main Section having last row of Side between markers on Main Section. With RS of Main Section facing, join yarn with sl st to lower edge. Ch 1. Work 8 sc through both thicknesses up side, 17 sc across top and 8 sc down other side. Do not turn.
Next row Ch 1. Working from left to right, intead of from right to left as usual, work 1 reverse sc in each sc to end of row. Fasten off.
Rep for other side.

SPA containers

YOU'LL NEED

YARN
Lily® Sugar 'n Cream Twists
2oz [56.7g], 95yds [87m]

1 ball 20116 (Country Twists) or 20117 (Denim Twists)

HOOK
Size D (3.25mm) crochet hook *or size to obtain gauge*

ADDITIONAL
Fabric stiffener (optional)

MEASUREMENTS
Approx 3" [7.5cm] in diameter x 4" [10cm] tall.

GAUGE
19 sc and 20 rows = 4" [10cm]. *Remember to check your gauge!*

CONTAINER
Ch 4. Join with sl st in ring.
1st rnd Ch 1. 8 sc in ring. Join with sl st in first sc.
2nd rnd Ch 1. 2 sc in each sc around. Join with sl st in first sc—16 sc.
3rd rnd Ch 1. *2 sc in next sc. 1 sc in next sc. Rep from * around. Join with sl st in first sc—24 sc.
4th rnd Ch 1. *2 sc in next sc. 1 sc in each of next 2 sc. Rep from * around. Join with sl st in first sc—32 sc.
5th rnd Ch 1. *2 sc in next sc. 1 sc in each of next 3 sc. Rep from * around. Join with sl st in first sc—40 sc.
6th rnd Ch 1. Working through back loops only, 1 sc in each sc around. Join with sl st in first sc.
7th rnd Ch 1. Working through both loops, 1 sc in each sc around. Join with sl st in first sc.
Rep last rnd 5 times more.
Next rnd Ch 1. 1 sc in each of first 9 sc. *2 sc in next sc. 1 sc in each of next 9 sc. Rep from * to last sc. 2 sc in last sc. Join with sl st in first sc—44 sc.
Work 6 rnds even.
Next rnd Ch 1. 1 sc in each of first 10 sc. *2 sc in next sc. 1 sc in each of next 10 sc. Rep from * to last sc. 2 sc in last sc. Join with sl st in first sc—48 sc.
Work 6 rnds even. Fasten off.

FINISHING
Optional Apply fabric stiffener to containers according to manufacturer's directions.

4TH OF July

YOU'LL NEED

YARN
Lily® Sugar 'n Cream Cotton 4 ply Solids 2.5oz [70.9 g], 120yds [110m]

1 ball each
Color A - 01 (White)
Color B - 09 (Bright Navy)
Color C - 95 (Red)

Note 1 ball makes 23 stars.
1 ball each of A, B and C makes 20 flags

HOOK
Size G (4.0mm) crochet hook *or size to obtain gauge*

ADDITIONAL
Glue gun and glue sticks
2¼yds [2m] Wrights® Gold Stars ⅞" [2cm] wide ribbon
Styrofoam wreath 12" [30.5cm] in diameter
1 ball Sugar'n Cream #09 Bright Navy to wrap wreath (optional)

GAUGE
16 sc and 16 rows = 4" [10cm]. *Remember to check your gauge!*

FLAG (make 4)
With C, ch 19.
1st row 1 sc in 2nd ch from hook. 1 sc in each ch to end of ch. Turn—18 sc.
2nd row With A, ch 1. 1 sc in back loop of each sc to end. Turn.
3rd row Ch 1. Sl st in back loop of each sc to end. Turn.
4th row With C, ch 1. 1 sc in back loop of each st to end. Turn.
5th row Ch 1. Sl st in back loop of each sc to end. Turn.
6th and 7th rows Rep 2nd and 3rd rows.
8th row With C, ch 1. 1 sc in back loop of each of next 11 sts, changing to B at end of 11th st. With B, 1 sc in back loop of each of last 7 sts. Turn.
9th row With B, ch 1. Sl st in back loop of each of first 7 sts. Change to C. Sl st in back loop of each of last 11 sts. Turn.
10th row With A, ch 1. 1 sc in back loop of each of first 11 sts. Change to B. 1 sc in back loop of each of last 7 sts. Turn.
11th row With B, ch 1. Sl st in back loop of each of first 7 sts. Change to A. Sl st in back loop of each of last 11 sts. Turn.
12th row With C, ch 1. 1 sc in back loop of each of first 11 sts. Change to B. 1 sc in back loop of each of last 7 sts. Turn.
13th row With B, ch 1. Sl st in back loop of each of first 7 sts. Change to C. Sl st in back loop of each of last 11 sts. Fasten off.

FINISHING
With A, embroider 8 French knots on Flag as illustrated (see diagram).

STAR (make 4 each in A, B and C)
Base
Ch 2.
1st rnd 10 sc in 2nd ch from hook. Join with sl st in first sc.
2nd rnd Ch 1. 2 sc in each sc around. 20 sc. Join with sl st in first sc.

First Point
****Next row** Ch 1. 1 sc in each of next 4 sc. Turn.
Next row Ch 1. 1 sc in each of next 4 sc. Turn.
Next row Ch 1. [(Draw up a loop in each of next 2 sc. Yoh and draw a loop through all loops on hook - Sc2tog made] twice. Turn.
Next row Ch 1. Sc2tog. Turn.
Ch 2. Sl st along side of Point to Base. Sl st in last sc worked on 2nd rnd.
Rep from ** 4 times more. 5 Points in total. Fasten off.

ASSEMBLY
If desired, wrap yarn around Styrofoam Wreath to cover completely. Glue Flags to Wreath, equally spaced, with slight curve or bend in the Flag, so that it appears to wave. Glue 3 Stars (one each A, B and C) in a cluster between Flags. Wrap ribbon loosely around Wreath and glue in position to back of Wreath.

FRENCH KNOT

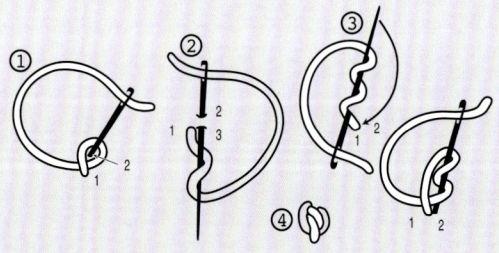

34

BEACH Bag

YOU'LL NEED

YARN
Lily® Sugar 'n Cream Cotton 4 ply Solids 2.5oz [70.9g], 120yds [110m]

5 balls Color A - 09 (Bright Navy)
2 balls Color B - 95 (Red)
1 ball Color C - 01 (White)

HOOK
Size 7 (4.5mm) crochet hook *or size to obtain gauge*

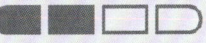

MEASUREMENTS
Finished size Approx 18" x 15" [45.5 x 38cm]

GAUGE
14 sts and 8 rows = 4" [10cm] in pat. *Remember to check your gauge!*

BAG

Front and Back (make 2 pieces alike) (crocheted sideways)
Note Ch 3 counts as first dc in row. Ch 2 does not count as first hdc in row.
With A, ch 51.
1st row 1 hdc in 3rd ch from hook. 1 hdc in each ch to end of ch. Turn—49 sts.
**2nd row Ch 3 (counts as 1 dc). Skip first hdc. 1 dc in each hdc to end of row. Turn.
3rd row Ch 2. 1 hdc in each dc to end of row. Turn.
4th row As 2nd row.
Last 2 rows form pat.**
Cont working in pat, in the following color sequence (always change color at the end of the previous row)
2 rows in B.
2 rows in C.
2 rows in B.
2 rows in C.
2 rows in B.
8 rows in A.
2 rows in B.
2 rows in C.
2 rows in B.
2 rows in C.
2 rows in B.
4 rows in A.
Fasten off.

Side/Bottom Panel (make 1)
With A, ch 18.
1st row 1 hdc in 3rd ch from hook. 1 hdc in each ch to end of ch. Turn—16 sts.
Rep from ** to ** as given for Front and Back.
Cont in pat until work from beg measures 48" [122cm].
Fasten off.

Straps (make 2)
With B, ch 113.
1st row 1 sc in 2nd ch from hook. 1 sc in each ch to end of ch. Turn—112 sts.
2nd to 4th rows Ch 1. 1 sc in each sc to end of row. Turn.
Fasten off.

FINISHING
Place WS of Front and Side/Bottom Panel tog and work 1 row of sc with A through all thicknesses. Rep for Back.

Edging
With RS facing, join A with sl st to top of Bag.
1st and 2nd rnds Ch 1. 1 sc in same st as last sl st. Work 1 rnd of sc around top of Bag. Sl st in first sc. Fasten off at end of 2nd rnd.
Sew Straps to Bag as shown in photo.
If desired, cut a piece of cardboard 5" x 17" [12.5 x 43cm] and place in bottom of Bag.

STARS and stripes cushion

YOU'LL NEED

YARN (4)
Lily® Sugar 'n Cream Cotton 4 ply Solids 2.5oz [70.9g], 120 yds [110m]

Four Cushions
3 balls Color A - 95 (Red)
3 balls Color B - 01 (White)
11 balls Color C - 09 (Bright Navy)

HOOK
Size 7 (4.5mm) crochet hook *or size to obtain gauge*

ADDITIONAL
Four 15" [38cm] square pieces of 1" [2.5cm] thick foam

MEASUREMENTS
Finished size Approx 15" x 15" [38 x 38cm]

GAUGE
14 dc and 7 rows = 4" [10cm]. *Remember to check your gauge!*

CUSHION
Front (make 4)
With A, ch 54.
1st row 1 dc in 4th ch from hook. 1 dc in each ch to end of ch. Turn—52 dc.
2nd row Ch 3 (counts as 1 dc). Skip first dc. 1 dc in each dc to end of row. Join B. Turn.
3rd and 4th rows With B, ch 3. Skip first dc. 1 dc in each dc to end of row. Join A. Turn.
5th and 6th rows With A, as 2nd row, changing to B at end of 6th row.
Rep last 4 rows once more, then 3rd and 4th rows once.
13th row With A, ch 3. Skip first dc. 1 dc in each of next 26 dc, changing to C at end of last dc. With C, 1 dc in each dc to end of row. Turn.
14th row With C, ch 3. Skip first dc. 1 dc in each of next 24 dc, changing to A at end of last dc. With A, 1 dc in each dc to end of row. Join B. Turn.
15th row With B, ch 3. Skip first dc. 1 dc in each of next 24 dc, changing to C at end of last dc. With C, 1 dc in each dc to end of row. Turn.
16th row With C, ch 3. Skip first dc. 1 dc in each of next 26 dc, changing to B at end of last dc. With B, 1 dc in each dc to end of row. Join A. Turn.
Rep last 4 rows twice more, then 13th and 14th rows once, omitting color change at end of 14th row. Fasten off.
With RS facing, join C with sl st to any corner of Front. Ch 1. Work 1 row of sc around Front, working 3 sc in corner sc's. Fasten off.
With B, embroider stars as illustrated.

Back (make 4)
With C, ch 54.
1st row 1 dc in 4th ch from hook. 1 dc in each ch to end of row. Turn—52 dc.
2nd row Ch 3 (counts as 1 dc). Skip first dc. 1 dc in each dc to end of row. Turn.
Cont in dc until Back measures the same as Front. Fasten off.
With RS facing, join C with sl st to any corner of Back. Ch 1. Work 1 row of sc around Back, working 3 sc in corner sc's. Fasten off.

FINISHING
With WS of Front and Back tog, join C with sl st to any corner. Working through both thicknesses, work 1 row of sc around 3 sides of Cushion, working 3 sc in corner sc's. Insert foam piece. Sc 4th side of Cushion closed. Sl st in first sc.

CHECKERED picnic blanket

YOU'LL NEED

YARN (4)
Lily® Sugar 'n Cream Cotton 4 ply Solids 2.5oz [70.9g], 120yds, [110m]

Blanket
13 balls Color A - 95 (Red)
13 balls Color B - 01 (White)
2 balls Color C - 09 (Bright Navy)

Coasters
2 balls Color B - 01 (White)
2 balls Color C - 09 (Bright Navy)

HOOK
Size 7 (4.5mm) crochet hook *or size to obtain gauge*

MEASUREMENTS
Finished size Approx 65" x 65" [164.5 x 164.5cm]

GAUGE
14 dc and 7 rows = 4" [10cm]. *Remember to check your gauge!*

BLANKET
Strip 1 (make 4)
With A, ch 30.
1st row 1 dc in 4th ch from hook. 1 dc in each ch to end of ch. Turn—28 dc.
2nd row Ch 3. Skip first dc. 1 dc in each dc to end of row. Turn.
Rep 2nd row 12 times more.
With B, rep 2nd row 14 times more.
These 28 rows form pat.
Cont in pat until work from beg measures approx 64" [162cm] ending on a 28th row of pat. Fasten off.

Strip 2 (make 4)
Rep as for Strip 1 substituting A for B and B for A.

FINISHING
Sew strips tog alternating Strip 1 and Strip 2 to form a checkerboard pattern.
Edging Join C with sl st, to bottom right corner of Blanket.
1st rnd Ch 1. 3 sc in same st as last sl st. Work 1 rnd of sc around Blanket, working 3 sc in each corner, 2 sc in each dc sp along sides and 1 sc in each sp between dc's along top and bottom. Join with sl st in first sc.
2nd rnd Ch 1. 1 sc in same sp as last sl st. 3 sc in next sc (corner). Proceed in sc around, working 3 sc in each corner sc. Join with sl st in first sc. Fasten off.

COASTERS (playing pieces) (make 12 each in B and C)
Ch 6. Join with sl st to form a ring.
1st rnd Ch 1. 10 sc into ring. Join with sl st in first sc.
2nd rnd Ch 1. 1 sc in same sp as last sl st. *2 sc in next sc. 1 sc in next sc. Rep from * around, working 2 sc in last sc. Join with sl st in first sc—15 sc.
3rd rnd As 2nd rnd, working 1 sc in last sc—22 sc.
4th rnd As 2nd rnd—33 sc.
5th rnd Ch 1. 1 sc in same sp as last sl st. 1 sc in each sc around. Sl st in first sc.
6th rnd As 3rd rnd—49 sc. Fasten off.

OVEN mitts

YOU'LL NEED

YARN (4)
Lily® Sugar 'n Cream Cotton 4 ply Solids 2.5oz [70.9g], 120yds [110m]

Two Mitts
2 balls Color A - 09 (Bright Navy)
1 ball Color B - 01 (White)

HOOK
Size 7 (4.5mm) crochet hook *or size to obtain gauge*

GAUGE
14 sc and 16 rows = 4" [10cm]. *Remember to check your gauge!*

FRONTS AND BACKS (make 4 pieces alike)
With A, ch 24.
1st row 1 sc in 2nd ch from hook. 1 sc in each ch to end of ch. Turn. 23 sts.
2nd row Ch 1. 1 sc in each sc to last 2 sc. 2 sc in next sc. 1 sc in last sc. Turn.
3rd to 6th rows Ch 1. 1 sc in each sc to end of row. Turn.
7th row Ch 1. Draw up a loop in each of next 2 sc. Yoh and draw through all loops on hook - Sc2tog made. 1 sc in each sc to end of row. Turn.
8th row Ch 1. 1 sc in each sc to last 5 sts. Turn. Leave rem sts unworked.
9th row Ch 13. 1 sc in 2nd ch from hook. 1 sc in each of rem 11 ch. 1 sc in each sc to end of row. Turn. 30 sts.
10th row Ch 1. 1 sc in each sc to last 2 sts. 2 sc in next sc. 1 sc in last sc. Turn.
11th row Ch 1. 2 sc in next sc. 1 sc in each sc to end of row. Turn.
12th to 14th rows Rep 10th and 11th rows once more, then 10th row once.
15th to 20th rows Ch 1. 1 sc in each sc to end of row. Turn.
21st row Ch 1. Sc2tog. 1 sc in each sc to end of row. Turn.
22nd row Ch 1. 1 sc in each sc to last 3 sts. Sc2tog. 1 sc in last sc. Turn.
23rd to 25th rows Rep 21st and 22nd rows once more, then 21st row once. 30 sts.
26th row Ch 1. 1 sc in each sc to end of row. Fasten off.

STAR (make 2)
Base
With B, ch 2.
1st rnd 10 sc in 2nd ch from hook. sl st in first sc.
2nd rnd Ch 1. 2 sc in each sc around. (20 sc). Sl st in first sc.
First Point: **Next row** Ch 1. 1 sc in each of next 4 sc. Turn.
Next row Ch 1. (Sc2tog) twice. Turn.
Next row Ch 1. Sc2tog. Turn.
Ch 2. Sl st along side of Point to Base.**
Second to Fifth Points Work from ** to ** as given for First Point.
Fasten off.

FINISHING
With WS facing, pin 2 Mitt pieces tog. Join A with sl st, to base of thumb section at side seam. Ch 1. Working through both thicknesses, work 1 row of sc around outer edge of Mitt to join 2 pieces tog, leaving bottom open. Work 1 rnd of sc around opening of Mitt. Ch 6 for hanging loop. Sl st in base of ch. Fasten off. Rep for second Mitt. Sew a Star to back of each Mitt as shown in image.

FLAG apron

YOU'LL NEED

YARN
Lily® Sugar 'n Cream Cotton 4 ply
Solids 2.5oz [70.9g], 120yds [110m]

4 balls Color A - 01 (White)
1 ball Color B - 09 (Bright Navy)
1 ball Color C - 95 (Red)

HOOK
Size 7 (4.5mm) crochet hook *or size to obtain gauge*

MEASUREMENTS
Finished size Approx 21" x 24" [53.5 x 61cm]

GAUGE
14 hdc and 11 rows = 4" [10cm]. *Remember to check your gauge!*

APRON
With A, ch 74.
1st row 1 hdc in 3rd ch from hook. 1 hdc in each ch to end of ch. Turn—72 sts.
2nd row Ch 2 (ch 2 does not count as first st). 1 hdc in each st to end of row. Turn. Rep last row until work from beg measures 16" [40.5cm].
Next row Ch 2. (Yoh. Draw up a loop in next st) twice. Yoh and draw through all loops on hook - Hdc2tog made. 1 hdc in each st to last 3 sts. Hdc2tog. 1 hdc in last st. Place markers at each end of row. Turn. Rep last row until 28 sts rem. Fasten off.

Edging
1st rnd With RS facing, join A with sl st to bottom right corner of Apron. Ch 1. 3 sc in same sp. Work 71 sc across bottom edge, 3 sc in corner st, 53 sc up side edge, 2 sc in marked st, 23 sc up armhole edge, 3 sc in corner st, 27 sc across top edge, 3 sc in corner st, 23 sc down armhole edge, 2 sc in marked st, 53 sc down side edge. Join with sl st in first sc. Fasten off.
2nd rnd Join B with sl st, to any corner. Ch 1. 3 sc in same st as last sl st. Work 1 sc in each sc around, working 3 sc in corner sts and 2 sc in marked sts. Join with sl st in first sc. Fasten off.

Neck Strap
With B, ch 71.
1st row 1 sc in 2nd ch from hook. 1 sc in each ch to end of ch. Turn.
2nd and 3rd rows Ch 1. 1 sc in each sc to end of row. Turn. Fasten off at end of 3rd row.

Ties (make 2)
With B, ch 101.
1st row 1 sc in 2nd ch from hook. 1 sc in each ch to end. Fasten off.

Flag Pocket With C, ch 48.
1st row (RS) 1 hdc in 3rd ch from hook. 1 hdc in each ch to end of ch. Turn—46 sts.
2nd row Ch 2. 1 hdc in each st to end of row, changing to A at end of last st. Turn.
3rd row With A, ch 2. 1 hdc in each st to end of row. Turn.
4th row Ch 2. 1 hdc in each st to end of row, changing to C at end of last st. Turn.
5th row With C, ch 2. 1 hdc in each st to end of row. Turn.
Rep 2nd to 5th rows once more, then 2nd to 4th rows once.
13th row With C, ch 2. 1 hdc in each of first 30 sts, changing to B at end of 30th st. With B, 1 hdc in each st to end of row. Turn.
14th row Ch 2. 1 hdc in each of first 16 sts, changing to C at end of 16th st. With C, 1 hdc in each st to end of row, changing to A at end of last st. Turn.
15th row With A, ch 2. 1 hdc in each of first 30 sts, changing to B at end of 30th st. With B, 1 hdc in each st to end of row. Turn.
16th row Ch 2. 1 hdc in each of first 16 sts, changing to A at end of 16th st. With A, 1 hdc in each st to end of row, changing to C at end of last st. Turn.
Rep 13th to 16th rows twice more, then 13th and 14th rows once, omitting color change at end of last row. Fasten off.

FINISHING
With A, embroider stars on Flag Pocket as illustrated. Sew Pocket to Apron. Sew along color change of Pocket through all thicknesses to form a smaller pocket alongside. Sew Neck Strap and Ties to Apron as illustrated.

TRICK or treat

YOU'LL NEED

YARN
Lily® Sugar 'n Cream Cotton 4 ply
Solids 2.5oz [70.9g], 120yds [110m]

1 ball each
Color A - 1131 (Celadon)
Color B - 82 (Jute)
Color C - 02 (Black)
Color D - 01 (White)
Color E - 1132 (Pumpkin)

Note 1 ball each of A, B and C makes 7 Witches
1 ball of D makes 10 Webs
1 ball of C makes 8 Spiders
1 ball of E makes 14 Small Pumpkins or 10 Large Pumpkins

HOOK
Size G (4.0mm) crochet hook *or size to obtain gauge*

ADDITIONAL
Glue gun and glue sticks
Grape Vine wreath 13" [33cm] in diameter
Fabric stiffener
6 glue-on eyes
Stuffing
Twigs for decoration

MEASUREMENT
Finished size Approx 13" diameter

GAUGE
16 sc and 16 rows = 4" [10cm]. *Remember to check your gauge!*

WITCH (make 2)
Cape
With C, ch 17.
1st row 1 sc in 2nd ch from hook. 1 sc in each ch to end of ch. Turn—16 sc.
2nd row Ch 1. Draw up a loop in each of next 2 sc. Yoh and draw a loop through all loops on hook - Sc2tog made. 1 sc in each sc to last 2 sc. Sc2tog. Turn.
3rd to 7th rows Rep 2nd row. 4 sc at end of 7th row. Fasten off.

Hat
With C, ch 10. Sl st to form ring.
1st rnd Ch 1. 1 sc in each ch around. Join with sl st to first st—10 sc.
2nd to 8th rnds Ch 1. Sc2tog. 1 sc in each sc to end of rnd. Join with sl st to first st.
9th rnd Ch 1. Draw up a loop in each of next 3 sc. Yoh and draw a loop through all loops on hook – Sc3tog made. Fasten off.

Brim
Join C with sl st to foundation ch of Hat. 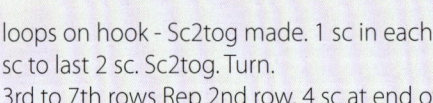 **1st rnd** Ch 1. 2 sc in each ch around. 20 sc. Join with sl st to first sc.
2nd rnd Ch 1. (1 sc in next sc. 2 sc in next sc) 10 times. Join with sl st to first sc. Fasten off.

Hair
Cut 30, 6" [15cm] lengths of A and tie securely in the middle.

Broom
Cut 12, 4" [10cm] lengths of B and tie securely in the middle.
Fold in half and tie tog 1" [2.5cm] down from fold. Glue to end of twig.

WEB (make 2)
With D, ch 8. Join with sl st to form ring.
1st rnd Ch 8. (Yoh) 3 times. Insert hook into ring. Yoh and draw up a loop through ring only. (Yoh and draw through 2 loops on hook) 4 times – Spoke made. Ch 4. (Work Spoke in ring. Ch 4) 6 times. Sl st in 4th ch of ch-8.
2nd rnd Ch 12. (Skip next 4 ch. Work Spoke in top of next Spoke. Ch 8) 7 times. Sl st in 4th ch of ch-12.
3rd rnd Ch 16. (Skip next 8 ch. Work Spoke in top of next Spoke. Ch 12) 7 times. Sl st in 4th ch of ch-16. Fasten off.

FINISHING
Stiffen Web with fabric stiffener.

SPIDER (make 3)
Body With C, ch 3.
1st rnd 8 hdc in 3rd ch from hook. Sl st in first hdc.
2nd rnd Ch 1. 2 sc in each hdc around. Sl st in first sc—16 sc.
3rd rnd Ch 1. 1 sc in back loop only of each sc around. Sl st in first sc.
4th rnd Ch 1. (Sc2tog) 8 times—8 sts.
Break yarn, leaving a long end. Thread yarn through needle and draw tightly through 8 sts. Fasten off.

Legs
Join yarn in front loop of any st in 3rd row with sl st.
*Ch 12.
Skip first ch. Sl st in each of next 11 ch. Sl st in same loop as initial join.
Sl st in each of next 2 loops of 3rd rnd. Rep from * 7 times more. Fasten off.

FINISHING
Glue 2 eyes on each Spider as in photo.

PUMPKIN (make 5 small and 1 large)
Small Pumpkin
With E, ch 4. Join with sl st to form ring.
1st rnd Ch 1. 8 sc in ring. Sl st to first sc.
2nd rnd Ch 1. 2 sc in each sc around. Sl st to first sc—16 sc.
3rd and 4th rnds Ch 1. 1 sc in each sc around. Sl st to first sc.
5th rnd Ch 1. (Sc2tog. 1 sc in each of next 2 sc) 4 times. Sl st in first st—12 sc.
Stuff with small amount of stuffing.
6th rnd Ch 1. (Sc2tog) 6 times. Sl st in first st—6 sc.
7th rnd Ch 1. (Sc2tog) 3 times. Fasten off (bottom of Pumpkin).

Large Pumpkin
With E, ch 4. Join with sl st to form ring.
1st rnd Ch 1. 8 sc in ring. Sl st to first sc.
2nd rnd Ch 1. 2 sc in each sc around. Sl st in first sc—16 sc.
3rd rnd Ch 1. (1 sc in next sc. 2 sc in next sc) 8 times. Sl st in first sc—24 sc.
4th and 5th rnds Ch 1. 1 sc in each sc around. Sl st in first sc.
6th rnd Ch 1. (Sc2tog. 1 sc in each of next 2 sc) 6 times. Sl st in first st—18 sc.
7th rnd Ch 1. (Sc2tog. 1 sc in next sc) 6 times. Sl st in first st—12 sc.
Stuff with small amount of stuffing.
8th rnd Ch 1. (Sc2tog) 6 times. Sl st in first st—6 sc.
9th rnd Ch 1. (Sc2tog) 3 times. Fasten off (bottom of Pumpkin).

Stem (make 1 for each Pumpkin)
With B, ch 3. Sl st in 2nd ch from hook. Sl st in last ch. Fasten off.

FINISHING
With E, sew through center of Pumpkin 5 times, each time bringing yarn around the body at evenly spaced intervals. Draw tightly and fasten securely. Sew Stem in place to foundation ch at top of Pumpkin.

Assembly
Using photo as a guide, glue stiffened Web to each side of Wreath as illustrated. Glue one Spider to each Web, making sure each Leg is glued in place to appear as if the Spider is walking. Glue the Witches on either side of the Wreath as illustrated, by layering first the Cape, then the Hair and then the Hat. Glue the Broom to the side of the Witch. Glue twigs and Pumpkin as illustrated. Glue a small length of white yarn to 3rd Spider and hang from top center of Wreath.

AUTUMN harvest

YOU'LL NEED

YARN
Lily® Sugar 'n Cream Cotton 4 ply Solids 2.5oz [70.9g], 120yds [110m] and Lily® Sugar'n Cream® Ombres 2oz [56.7g], 95yds [87m]

1 ball each
Color A - 1130 (Warm Brown)
Color B - 101 (Soft Gold)
Color C - 2234 (Country Brown) (Ombre)
Color D - 82 Jute

Note 1 ball Solid makes 7 Leaves
1 ball Ombre makes 6 Leaves
1 ball each of Solids A and D makes 23 Acorns

HOOK
Size G (4.0mm) crochet hook *or size to obtain gauge*

ADDITIONAL
Small amount of stuffing for acorns.
Glue gun and glue sticks.
Grape vine wreath 13" [33cm] in diameter.
Dried red berries on stems

MEASUREMENT
Finished size Approx 13" diameter

GAUGE
16 sc and 16 rows = 4" [10cm]. *Remember to check your gauge!*

LEAF
(Make 4 each in A and C. Make 2 in B)
Ch 2. **1st row** 1 sc in 2nd ch from hook. Turn. **2nd row** Ch 1. 1 sc in sc. Turn. **3rd row** Ch 1. 2 sc in sc. Turn. **4th row** Ch 1. 2 sc in each of next 2 sc. Turn—4 sc. **5th to 8th rows** Ch 1. 2 sc in first sc. 1 sc in each sc to last sc. 2 sc in last sc. Turn—12 sc at end of 8th row. **9th row** Ch 1. 1 sc in each sc to end of row. Turn. **10th row** 1 hdc in first st. 1 dc in next st. 1 hdc in next st. Sl st in next st. 1 sc in each of next 4 sts. Sl st in next st. 1 sc in each of last 3 sts. Turn. **11th row** Ch 1. Draw up a loop in each of next 2 sts. Yoh and draw a loop through all loops on hook - Sc2tog made. 1 sc in next st. Sl st in next st. 1 sc in each of next 4 sts. Turn. Leave rem sts unworked. **12th row** Ch 1. 1 sc in each of first 4 sts. Turn. Leave rem sts unworked. **13th to 16th rows** Ch 1. 2 sc in first sc. 1 sc in each sc to last sc. 2 sc in last sc. Turn—12 sc at end of 16th row. **17th to 19th rows** Rep 10th to 12th rows. **20th row** Ch 1. (Sc2tog) twice. Turn. **21st row** Ch 1. Sc2tog. Fasten off.

Edging
With RS facing, join yarn at stem end. Ch 1. Sc evenly around, working 3 sc in each outside curve and working 2 sl st at each inside corner.

Stem
With RS facing, join yarn to ch at base of stem end of Leaf. Ch 6. Sl st in 2nd ch from hook. Sl st in each ch across. Sl st in sc at base of Stem. Fasten off.

FINISHING
Fold Leaf in half lengthways, with RS tog. Sew a seam ¼" [5 mm] from fold, along center 3½" [9cm] of Leaf.

ACORN (make 9)
Acorn Body
With D, ch 2. **1st rnd** 8 sc in 2nd ch from hook. Join with sl st in first sc. **2nd rnd** Ch 1. 2 sc in each sc around. Join with sl st in first sc. 16 sc. **3rd rnd** Ch 1. 1 sc in each sc around. Join with sl st in first sc. **4th rnd** Ch 1. (Sc2tog) 8 times. Join with sl st in first sc. 8 sc. Fasten off. Stuff lightly.

Acorn Cap
With A, ch 2. **1st rnd (RS)** 8 sc in 2nd ch from hook. Join with sl st in first sc. **2nd rnd** Ch 1. 2 sc in each sc around. Join with sl st in first sc—16 sc. **3rd rnd** Ch 1. 1 sc in each sc around. Join with sl st in first sc. Fasten off, leaving a long end for sewing Cap to Body.

Stem
With A, ch 4.
Sl st in 2nd ch from hook. Sl st in each of next 2 ch. Fasten off.

FINISHING
With WS of Cap showing, sew Acorn Cap to top of Acorn Body. Attach Stem.

Assembly
Using photo as a guide, glue Leaves to Wreath as illustrated. Glue on Acorns in clusters as illustrated. Glue on dried berries.

SEASONS greetings

YOU'LL NEED

YARN
Lily® Sugar 'n Cream Cotton 4 ply Solids 2.5oz [70.9g], 120yds [110m]

1 ball each
Color A - 16 (Dark Pine)
Color B - 960 (White Tinsel)

Note 1 ball of A makes 35 holly leaves
1 ball of B makes 15 bells

HOOK
Size G (4.0mm) crochet hook *or size to obtain gauge*

ADDITIONAL
Gold beads ¼" [10mm] diameter
Brass coated wire ½yd [.5m] long
1½yds [1.4m] white ribbon with gold trim 1" [2.5cm] wide
1yd [.9m] of red ribbon ⅜" [12mm] wide
8 small dried pine cones
Glue gun and glue sticks
Grape Vine wreath 13" [33cm] in diameter

MEASUREMENT
Finished size Approx 13" diameter

GAUGE
16 sc and 16 rows = 4" [10cm]. *Remember to check your gauge!*

BELL (make 15)
With B, ch 2.
1st rnd (RS) 6 sc in 2nd ch from hook. Join with sl st in first sc.
2nd rnd Ch 1. 1 sc in each sc around. Join with sl st to first sc.
3rd rnd Ch 1. 1 sc in each of first 2 sc. 2 sc in next sc. 1 sc in each of next 2 sc. 2 sc in next sc. Join with sl st in first sc—8 sc.
4th rnd Ch 1. 1 sc in same st as last sl st. (2 sc in next sc. 1 sc in next sc) 3 times. 2 sc in next sc. Join with sl st in first sc—12 sc.
5th to 7th rnds Ch 1. 1 sc in same st as last sl st. 1 sc in each sc around. Join with sl st in first sc. Fasten off at end of 7th rnd.

FINISHING
Cut desired length of wire and glue bead at one end. Insert other end of wire into Bell and glue it in position as illustrated.

HOLLY LEAF (make 15)
With A, ch 11.
1st rnd (RS) 1 sc in 2nd ch from hook. 1 sc in next ch. (1 hdc. Ch 2. 1 hdc) all in next ch. 1 hdc in each of next 2 ch. (1 dc. Ch 2. 1 dc) all in next ch. 1 dc in next ch. 1 hdc in each of next 2 ch. (1 sc. Ch 3. 1 sc) all in last ch.

Working in other side of ch, proceed as follows

1 hdc in each of next 2 ch. 1 dc in next ch. (1 dc. Ch 2. 1 dc) all in next ch. 1 hdc in each of next 2 ch. (1 hdc. Ch 2. 1 hdc) all in next ch. 1 sc in next ch. 1 sc in last ch. Ch 2. Join with sl st to first sc. Fasten off.

ASSEMBLY
Using photo as a guide, glue Leaves, Bells and small bows in clusters as illustrated. Glue clusters on Wreath. Glue big bow at top of Wreath as in photo. Glue pine cones onto Wreath as in photo.

MERRY Christmas

YOU'LL NEED

YARN 4
Lily® Sugar 'n Cream Cotton 4 ply
Solids 2.5oz [70.9g], 120yds [110m]

1 ball each
Color A - 95 (Red)
Color B - 47 (Baby Pink)
Color C - 01 (White)
Color D - 800 (Mistletoe Sparkle)
Small amount of blue yarn

Note 1 ball each of A, B and C makes
15 Santas
1 ball each of A and C makes
7 Stockings
1 ball of Sparkle makes 5 Gift Boxes

1 ball Sugar'n Cream #62 Emerald to wrap Wreath (optional)

HOOK
Size G (4.0mm) crochet hook *or size to obtain gauge*

ADDITIONAL
Stuffing
Glue gun and glue sticks
2yds [1.8m] of ⅝" [15mm] wide green ribbon
3yds [2.75m] of ¼" [10mm] wide white ribbon
20 small dried pine cones
Styrofoam Wreath 12" [30.5cm] in diameter

MEASUREMENT
Finished size Approx 12" diameter

GAUGE
16 sc and 16 rows = 4" [10cm]. *Remember to check your gauge!*

SANTA (make 3)
Face
With B, ch 6. Join with sl st to form ring.
1st rnd Ch 3. 7 dc in ring. Break B and join C in last dc. With C, 8 dc in same ring. Sl st in first dc.
2nd rnd Ch 3. Sl st in top of first B st. *Ch 3. Sl st in top of next st.* Rep from * to * for all B sts. **Ch 5. Sl st into next C st.** Rep from ** to ** to complete ring. Fasten off.

Hat
With A, ch 9.
1st row 1 sc in 2nd ch from hook. 1 sc in each sc to end of ch—8 sc. Turn.
2nd row Ch 1. Skip first sc. 1 sc in each sc to end of row. Turn.
Rep 2nd row until one st remains. Fasten off.

Pompom
Join C with sl st to point of Hat. Ch 4. Join with sl st to form ring.
1st rnd Ch 1—12 sc in ring.
Break yarn leaving 6" [15cm] end. Thread yarn through a needle and draw through top of each st in ring. Draw up firmly and fasten off.

FINISHING
Sew Hat to Face as illustrated. Embroider two eyes using blue yarn.

STOCKING (make 4)
Cuff
With C, ch 13.
1st row (RS) 1 sc in 2nd ch from hook. 1 sc in each ch to end of ch—12 sc. Ch 1. Turn.
2nd row 1 sc in each sc to end of row. Ch 1. Turn.
Rep 2nd row once more, joining A at end of last row.

Leg
With A, 1 sc in each sc to end of row. Ch 1. Turn.
Rep last row until work from beg measures 3" [7.5cm] ending with RS facing and omitting turning ch at end of last row. Fasten off.

Heel
With RS of work facing, skip first 9 sc. Join C with sl st in next sc. Ch 1. 1 sc in same sc. 1 sc in each of last 2 sc. Bring other side of leg around. Work 1 sc in each of first 3 sc. Ch 1. Turn.
Next row 1 sc in each of first 4 sc. Ch 1. Turn.
Next row 1 sc in each of first 2 sc. Ch 1. Turn.

Next row 1 sc in each of next 2 sc. 1 sc in next sc of long row below. 3 sts. Ch 1. Turn.
Next row 1 sc in each of next 3 sc. 1 sc in next sc of long row below. 4 sts. Ch 1. Turn.
Cont as for last 2 rows until 6 sts of Heel have been worked, omitting turning ch at end of last row. Fasten off.

Foot
With RS of work facing, skip first 3 sc of Heel. Join A with sl st in next sc. Ch 1. 1 sc in same sc. 1 sc in each of next 2 sc of Heel. Skip next sc of Leg. 1 sc in each of next 4 sc. Skip next sc of Leg. 1 sc in each of next 3 sc of Heel. Ch 1. Turn—10 sts.
Next row 1 sc in each sc to end of row. Ch 1. Turn.
Next row 1 sc in each sc to end of row, changing to C in last sc. Ch 1. Turn.

Toe
Next 2 rows With C, 1 sc in each sc to end of row. Ch 1. Turn.
Next row (Draw up a loop in each of next 2 sc. Yoh and draw a loop through all loops on hook - Sc2tog made) 5 times. Fasten off.

FINISHING
Sew Cuff and Leg back seams. Sew Foot and Toe seams. Sew corners of Heel closed. Stuff lightly.

GIFT BOX (make 5)
Box Main Piece
With D, ch 8.
1st row 1 sc in 2nd ch from hook. 1 sc in each ch to end of ch. Turn—7 sc.
**2nd to 6th rows Ch 1. 1 sc in each sc to end of row. Turn.
7th row Ch 1. 1 sc in back loop only of each sc to end of row. Turn.
8th to 10th rows Ch 1. 1 sc in each sc to end of row. Turn.**
11th row Rep 7th row.
Rep from ** to ** once. Fasten off.

Box Sides (make 2)
Ch 7.
1st row 1 sc in 2nd ch from hook. 1 sc in each ch to end of row. Turn. 6 sc.
2nd to 4th rows Ch 1. 1 sc in each sc to end of row. Turn.

FINISHING
Sew first row of Main Piece to last row to form a box with sides open. Sew one Box Side in position. Stuff Box. Sew remaining Box Side in position.
Tie a length of ribbon around Box, tie a bow, and stitch in place.

Assembly
If desired, wrap yarn around Styrofoam Wreath to cover completely.
Using photo as a guide, glue Santas, Stockings and Gift Boxes on Wreath as in photo. Glue on pine cones and ribbon bow.

SLIPknot

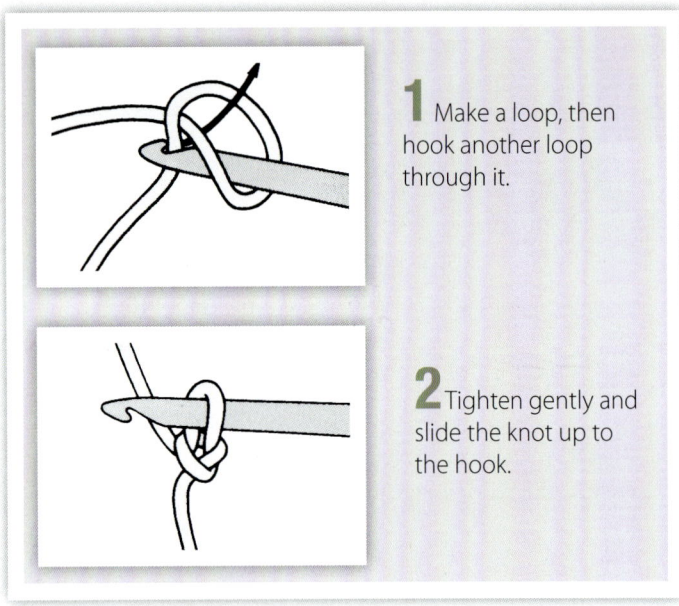

1 Make a loop, then hook another loop through it.

2 Tighten gently and slide the knot up to the hook.

CHAINstitch

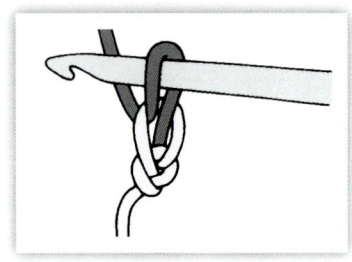

1 Yarn over hook (yo) and draw the yarn through to form a new loop without tightening up the previous one.

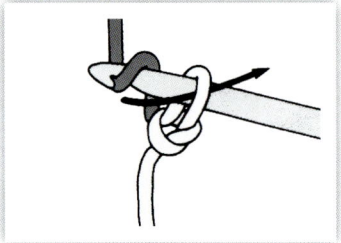

2 Repeat to form as many chain (ch) as required. Do not count the slip knot as a stitch.

SLIPstitch (sl st)

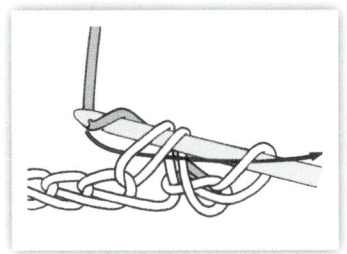
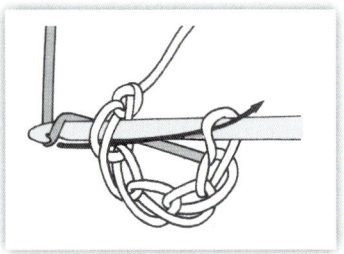

This is the shortest crochet stitch and unlike other stitches is not used on its own to produce a fabric. It is used for joining, shaping and where necessary carrying the yarn to another part of the fabric for the next stage.

To join a chain ring with a slip stitch (sl st), insert hook into first chain (ch), yarn over hook (yo) and draw through both the work and the yarn on hook in one movement.

Insert hook into work (second chain from hook), yarn over hook (yo) and draw the yarn through both the work and loop on hook in one movement.

SINGLEcrochet (sc)

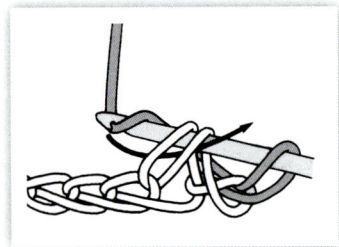
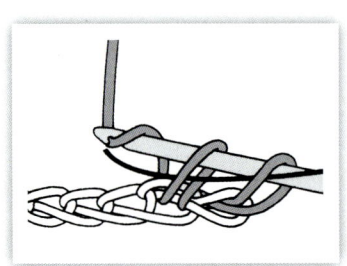

1 Insert the hook into the work (2nd chain (ch) from hook on starting chain), *yarn over hook (yo) and draw yarn through the work only.

2 Yarn over hook (yo) again and draw the yarn through both loops on the hook.

3 1 single crochet (sc) made. Insert hook into next stitch: repeat (rep) from * in step 1.

HALFdouble crochet

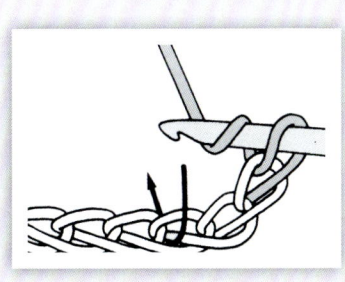

1 Yarn over hook (yo) and insert the hook into the work (3rd chain (ch) from hook on starting chain).

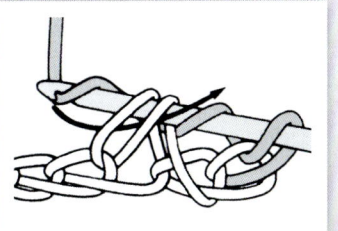

2 Yarn over hook (yo) and draw through the work only.

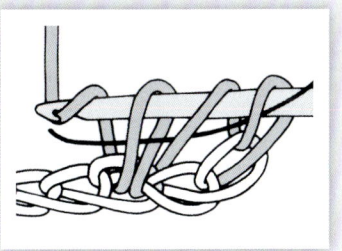

3 Yarn over hook (yo) again and draw through all three loops on the hook.

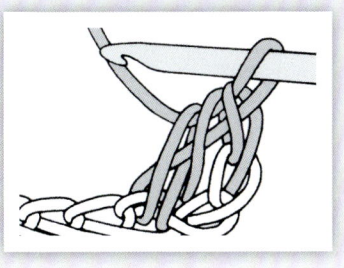

4 1 hdc made. Yarn over hook (yo), insert hook into next stitch (st); repeat (rep) from step 2.

DOUBLE crochet

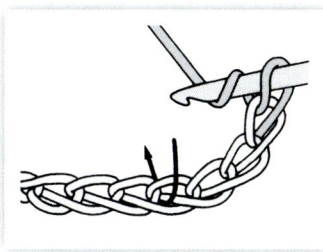

1 Yarn over hook (yo) and insert the hook into the work (4th chain from hook on starting chain).

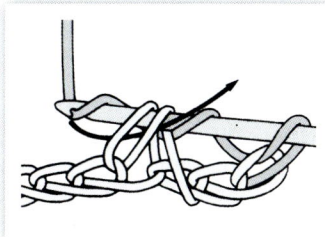

2 Yarn over hook (yo) and draw through the work only.

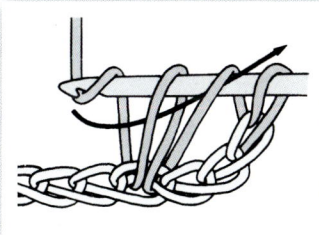

3 Yarn over hook (yo) and draw through the first two loops only.

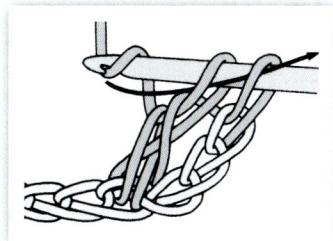

4 Yarn over hook (yo) and draw through the last two loops on the hook

5 1 dc made. Yarn over hook (yo), insert hook into next stitch (st), repeat (rep) from step 2.

NOTES